ANOVA SOUS VIDE COOKBOOK

LEGAL NOTICE

This book is copyright (c) 2018 by Ingrid Eakon. All rights are reserved. This book may not be duplicated or copied, either in whole or in part, via any means including any electronic form of duplication such as recording or transcription. The contents of this book may not be transmitted, stored in any retrieval system, or copied in any other manner regardless of whether use is public or private without express prior permission of the publisher.

This book provides information only. The author does not offer any specific advice, including medical advice, nor does the author suggest the reader or any other person engage in any particular course of conduct in any specific situation. This book is not intended to be used as a substitute for any professional advice, medical or of any other variety. The reader accepts sole responsibility for how he or she uses the information contained in this book. Under no circumstances will the publisher or the author be held liable for damages of any kind arising either directly or indirectly from any information contained in this book.

Anova Sous Vide Cookbook

100 Thermal Immersion Circulator Recipes for Precision Cooking At Home

By: Ingrid Eakon

Contents

Introduction 6

Chapter 1: Beef Recipes 17

Chapter 2: Lamb Recipe 35

Chapter 3: Chicken Recipe 53

Chapter 4: Pork Recipes 71

Chapter 5: Seafood Recipes 89

Chapter 6: Vegetable Recipes 107

Chapter 7: Dessert Recipes 125

The "Dirty Dozen" and "Clean 15" 136

Measurement Conversion Tables 137

Cooking Times and Temperatures 138

Recipe Index 141

Complete Index 143

Introduction

What is *Sous Vide*?

Sous Vide is a French term for a cooking method, literally translating to "under vacuum", which is also a very descriptive term for the cooking method. Cooking *Sous vide* usually requires food to be placed in vacuum sealed containers, such as a plastic Ziploc bag, or even in a glass Mason jar, and cooked by placing the container in a temperature controlled water or steam bath. The food is then usually cooked for a longer time than it usually takes as compared to conventional cooking methods, but the temperature is also set to be lower than conventional cooking methods. This method of cooking allows the food to be cooked evenly, without any risk of the outer portion of the food becoming overcooked while the inner portion of the food is being brought up to the requisite temperature. *Sous vide* also allows food to retain its natural moisture, as the food is kept sealed, minimizing water loss.

Sous vide cooking can be summarized into having three main components: first, the low cooking temperature of the method, second, the food being placed into sealed containers that prevent direct contact between the food and the heating method, and third, the pressurized containers, either through a full vacuum seal, or a simple partial vacuum.

How does the Sous Vide cooking method work?

The first step in employing the *sous vide* cooking method is to place the food that is to be cooked in a container. The container has to be thin enough to allow heat to be adequately transferred, and of a material that will hold up to the cooking temperatures used. Once the food is placed into the container, it should be sealed shut and if the cook so wishes, can be vacuumed to allow for the food to be pasteurized and even have a better consistency, as is the case with vegetables. This vacuum sealing also allows for quicker and more efficient cooking, as the food can be brought up to the requisite temperature quicker, as the air around it does not have to be heated first.

When the food is sealed and placed in the *sous vide* machine, the temperature of the water or steam bath should already have been set to the proper level in order to cook the food. This method ensures that the food will never become overcooked, following the simple laws of physics, as the food will never be at a higher temperature than the environment it is placed in, allowing the outer portion of the food to maintain a specific temperature as the inner portion cooks and eventually matches the temperature. This is in stark contrast to other cooking methods that make use of high temperatures, such as grilling or frying, as in these methods, food is exposed to temperatures far higher than the desired final cooking temperature, and this means that there is a risk of the outer portion of the food becoming overcooked and the inner portion being undercooked, as there is no way to ensure an even distribution of heat. This precision control of temperature is one of the major advantages of *sous vide*, and can be seen through the ease of cooking for thick cuts of meat, or food items that are not evenly shaped, as the long and slow even cooking method will result in perfectly cooked food given enough time.

The low temperature method of *sous vide* is also one of the crucial elements in creating an exceptional dish, as *sous vide* tends to result in much juicer foods as compared to standard cooking methods. This is because the low temperature of cooking means that the cell walls of the food items do not rupture, creating a more succulent result. When cooking meat products, the collagen contained in the tendons, the connective tissue of meat is hydrolyzed into gel, softening it. However, with the low temperature of the *sous vide* method, the proteins in the meat are not heated to the point that they denature, preventing them from becoming too tough and dry, allowing the cut of meat to retain its tenderness and moisture as compared to standard high heat methods. For vegetable items, the *sous vide* technique allows them to be cooked at a point below a hundred Celsius (100 C), the boiling point of water, allowing them to retain their texture, preventing them from becoming too soft. The *sous vide* method allows the polysaccharides contained in the vegetables to become de polymerized, as well as the gelatinization of the starches contained in the vegetables without the cell walls rupturing, leading to the retention of the vegetable's texture.

The vacuuming of the container is not necessary to the food becoming cooked, but it allows for much easier storage, as a vacuumed container will lead to the pasteurization of the food inside, giving it a much longer storage life. This vacuuming removes the food from possible exposure to oxygen, preventing food from oxidizing and thus keeping it from spoiling. It also helps when cooking certain foodstuffs, due to the longer period of time required for cooking food *sous vide*, as some foods may become rancid if allowed to oxidize, such as the fat in animal products.

Is *Sous Vide* Safe?

There may be some concerns when cooking with the Sous Vide method, especially when working with plastic, as some kinds of plastic have been known to release harmful chemicals if heated. However, food grade plastics, which are plastics made of high or low density polyethylene, are safe even when heated, and all plastic bags made specifically for Sous vide cooking are safe to be used. Even other non-specialized plastics are safe, such as Ziploc bags, but make sure to use food grade plastic containers, and avoid plastics made of polyvinyl chloride, as these are the types of plastics that lead to the release of harmful chemicals if heated. If in doubt, make sure to check what the plastic bag is made of, or you can simply choose to avoid the issue altogether and make use of a glass container.

As to those who worry about the safety of the food due to the low cooking temperature, due to the conventional wisdom of high temperature being needed to kill harmful bacteria, the length of time also plays a factor. Temperatures that are usually thought to be too low to make food safe can be sufficient to kill harmful bacteria if they are kept for a long enough period of time. One example of this would be some recipes for fish using a sous vide style that have a final cooking temperature of fifty five Celsius or a hundred and thirty one Fahrenheit (55 C / 131 F). However, if a person has a higher risk factor, such as if their immune system is compromised or they are pregnant, they may wish to avoid eating food that has not been cooked at a temperature sufficient to pasteurize them. Improperly cooked or uncooked food that is not exposed to oxygen may carry the Clostridium bacteria, and thus sous vide cooking should be done properly in order to avoid any botulism incidents. For most food, heating them to a temperature of at least fifty five Celsius or a hundred and thirty one Fahrenheit (55 C / 131 F) and keeping that for four hours should be sufficient for pasteurization. However, there is always a risk that some spores may survive and grow again once the food item cools, and in order to avoid that, the food can be chilled quickly, to a temperature of three Celsius / thirty seven Fahrenheit (3 C / 37 F) or lower when it is put into storage, to prevent the botulism bacteria from having the proper temperature to propagate.

As long as sous vide is done properly and with the correct materials, there is nothing to fear from the cooking method, and it may even prove to be safer and more effective in killing bacteria as compared to standard, conventional high heat cooking methods.

Does cooking food *Sous Vide* lead to better meals?

The exact temperature control of the *sous vide* cooking method creates unique results, leading to flavors and textures that conventional cooking techniques cannot replicate. The reason is simple, as the nature of exact temperature control is something that high heat conventional methods can never be able to achieve. Due to this, being able to consistently cook properly cooked food that tastes good is something that takes much more effort when done using high heat cooking methods, due to the balancing act that has to be done to ensure that the outside of the food is not overcooked, and to ensure that the inside of the food is properly cooked. The chance of food ending up dry, overcooked, and flavorless is always there, especially if the cook is inexperienced, but this can also happen to the best chefs.

The precision time and temperature of *sous vide* leads to consistency in food quality. The exact temperature and exact time will lead to the same result every time you cook, meaning that if you finally find a recipe that you enjoy, you can be sure that you will be able to replicate it over and over, with almost no risk of over or under cooking your food.

The sealing of the food in plastic or glass packaging allows the food to cook in its own juices, as well as combined with the low temperature, prevents the food from drying out. This allows food to reach a level of succulence that traditional methods of cooking cannot replicate, or will have an extremely difficult time doing. Food also does not shrink as much when cooked *sous vide* due to it not drying out, allowing for more food volume, and more food available every time you cook.

One of the most notable advantages of cooking *sous vide* is the flexibility that it affords the chef or cook. High heat conventional methods of cooking need constant attention to ensure that food gets properly cooked, but the *sous vide* method allows the food to reach a precise heat level and keeps it that way, allowing the cook to do other things in the meantime without fear of overcooking the food. It also allows less experienced chefs to create better dishes, as unlike with high heat methods of cooking, *sous vide* is far more forgiving when it comes to being able to cook a food item through without burning it or leaving it undercooked.

Some foods can be cooked through the *sous vide* and finished through another method of cooking, allowing it to become cooked thoroughly while maintaining distinctive flavors that are achieved through high heat methods. One example would be a steak: cooking it *sous vide* in order to cook it to a medium rare point all throughout the steak, ensuring it will be cooked evenly, then placing it on a grill or cast iron pan to sear the outside and impart that burnt crust is a great way to integrate multiple cooking techniques. Another standard food item that can be improved using the *sous vide* method is the simple egg. Whether the cook prefers their eggs to be poached, soft boiled, or hard boiled, cooking the egg at the exact temperature *sous vide* allows the cook to determine the exact consistency of the final product, something that can only be done with much difficulty and experience if using traditional methods. Some fish can also benefit from being cooked *sous vide*, as fish cooks quickly, so the traditional method of pan frying or grilling often leads to overcooked exteriors, something that can be eliminated entirely using the *sous vide* method of cooking.

How do *Sous Vide* Ovens / Baths usually work?

Sous vide ovens or baths tend to be large machines that contain a water bath or steam bath, and are operated through plugging them into an electric socket. These machines are able to control the temperature within the water bath, allowing them to be used for *sous vide* cooking. They are often controlled through an electronics panel found on the exterior of the machine, and must usually be filled to the brim with water in order to work properly, and these machines usually have a capacity of eleven litres (11 L), which may lead to wastage if cooking in small batches.

How does the Anova Precision Cooker differ from conventional *Sous Vide* machines?

The Anova Sous Vide Precision Cooker is a type of *sous vide* machine known as an immersion circulator. The biggest difference between an immersion circulator and a standard *Sous vide* machine is that an immersion circulator does not come with a built in water bath. Instead, it is meant to be attached to a pot full of water, and works by heating the water contained within the pot and allowing it to circulate, replicating the effect of a water bath, and allowing the maintenance of an exact temperature for *sous vide* cooking to be achieved. This means that immersion circulators make use of standard kitchenware to create the requisite water bath, rather than integrating a water bath into the device.

This has the notable advantage of being much smaller and easier to store as compared to a standard *sous vide* machine, as well as it being cheaper, as the Anova Precision Cooker retails at a price of ninety nine dollars (USD 99) for the eight hundred watt (800 W) Bluetooth capable version, and a hundred and twenty nine dollars (USD 129) for the nine hundred watt (900 W) Bluetooth and wi fi capable version. The 800 watt version allows the device to heat up enough water to cook about seven to eight servings, and the 900 watt version allows for about ten to twelve servings to be cooked at once. If the pot is too large for the device, the device may not be able to bring all the water in the pot to the proper temperature, and thus not be able to cook the food properly. The Bluetooth and wi fi capability on the devices also allows the user to control the device without physically adjusting the controls on the device, allowing for even greater flexibility in its usage.

Using the Anova *Sous Vide* Precision Cooker

One of the best things about using the Anova precision cooker is its ease of use. In order to begin cooking *sous vide*, all that has to be done is to clip the Anova precision cooker to a pot, plug in the Anova precision cooker, and fill the pot with water above the fill line of the device. Preheat the water by inputting the desired temperature in the Anova device, either by using the physical buttons, or through the Anova software. Once the play button is pressed, the Anova will begin to heat the water, and beep once the water has reached the requisite temperature. If you aren't there to hear the beep, a notification will also appear on the application.

Once the water is ready, simply clip or place your bagged food inside the pot (make sure that the bag is fully submerged. If it is likely to float, it may be clipped to the pot, or a weight may be placed on the edge to keep it from floating), and keep it there for the requisite amount of time in order for it to be properly cooked. The Anova can be set to automatically turn off after a certain amount of time, either through the physical interface or through the application. It can also be manually turned off if the user wishes. Once the food is done, it may already be eaten or stored, or if the cook wishes, they can give their food another layer of flavor by "finishing" it with another cooking method, such as pan searing, broiling, or grilling.

Tips and tricks for using the Anova Sous Vide Precision Cooker:

One of the characteristics of the *sous vide* cooking method is the vacuum sealing of the food in its cooking package. As earlier discussed, the vacuum sealing, while not precisely required, allows for faster cooking, and may also lead to better textures in the food. Vacuum sealing can be done through various methods. One method is to use a custom vacuum sealer and its specific bags, and use these in order to package the food you wish to cook *sous vide*. Take note that when vacuuming using this method, it may be possible to over vacuum the food, compressing it and ruining its texture. This possibility is most likely with delicate fish, as putting too much pressure may lead to it losing its flaky texture.

Another method is to use a pot of water and a Ziploc bag in order to create a vacuum. This can be done by placing the food in the bag, and slowly lowering the bag into the water, allowing the water pressure to force the air out of the bag. Once the water line has almost reached the top of the bag, simply seal it to create a vacuum. One more method is to use a Ziploc bag and simply suck the air out using a straw, though this method should not be used when handling raw meat, but can be employed when the food involved is fruit or vegetables. Glass jars are also recommended when cooking foods such as beans or grains, or even desserts. Do not worry about the use of plastic when cooking foods *sous vide*, as it is recommended to re use cooking containers when using the *sous vide* method. In fact, containers and packaging specifically designed for *sous vide* are re usable, as well as most other food grade plastics such as Ziploc bags.

Sous vide requires less technical expertise as compared to other cooking methods, and mastering cooking *sous vide* means mastering and learning the correct time and temperature for your food. These are the two main factors that the cook must keep in mind when cooking *sous vide*, as they are the only things that the cook can control to change how the food is cooked. Take note that though the temperature is controlled, keeping certain foods at a certain temperature for too long a time may have certain effects, such as for meats, which begin to break down if kept at a cooking temperature for too long – usually over four hours. However, the cook must also make sure that the food is cooked long enough in order to get rid of all the harmful bacteria that may be contained in the food.

While *sous vide* allows food to be cooked fully in the water bath, some foods are still better when exposed to some level of high heat. In order to add another dimension of flavor to the food, it can easily be finished using standard cooking techniques. A steak cooked *sous vide* can still have that classic sear in a few minutes if the cook wants to finish it using a cast iron pan. Some foods such as ribs or chops can be slapped on the grill to give it that slightly charred exterior. Even chicken cooked *sous vide* or vegetables can be broiled to give it an extra layer of flavor. Some starchy foods can even be fried, such as French fried potatoes initially cooked *sous vide*, or chicken cooked *sous vide*, then breaded and fried to create fried chicken. Take note that as the water in the *sous vide* bath does not come into any contact with food, it can be reused several times to cook food *sous vide*, or can be used for other purposes such as washing dishes, watering plants, or even giving to your pet to drink (once the water has cooled down), minimizing waste.

Keeping food simple when in the *sous vide* bath is often a better choice than including too many ingredients in one bag. Including dry herbs or a small amount of butter or oil in the food packaging will allow the food to absorb flavor while allowing its own natural flavors to come out. Take note that salting the food before cooking it *sous vide* may have an undesired effect, especially with meat, as salting may cure the meat and begin to break it down, drying out the meat. Some herbs are also far more potent than others, and proportions used when cooking using conventional methods may not have the same effect as when used in *sous vide* cooking, due to the extended period of time that the food spends cooking.

Putting too many liquids into the food package may result in the food not cooking properly, or taking too long to cook, due to the nature of the heating process. Too much liquid in the sealed packaging may even hurt the quality of the food, especially if the liquid has alcohol, or is high in acidity. This is because the sealed packaging does not allow for the reduction of the liquid while cooking, and there is a possibility that the liquid will take on bad flavors. Putting in marinating liquid during the cooking process may also not produce the desired result, as marinating liquids are usually meant to impart flavors to raw meats, and as the *sous vide* bath will begin to bring up the exterior to the final temperature rather quickly, the marinating liquid will not be able to penetrate the meat. However, marinating the meat beforehand, allowing the flavors to be absorbed, then draining the marinade and placing the meat in the sealed packaging can be done.

When using the Anova Sous Vide Precision Cooker, it may be helpful to cover the pot of water with wrap when cooking. This minimizes the amount of water lost during the cooking time, as the evaporated water will accumulate on the wrap and drip back down to the pot. This also has the advantage of keeping your Anova Precision Cooker safe, as it reduces the chance of moisture collecting in the electronic head of the device due to evaporate, and will prevent damage to the device.

How do I take care of my Anova Sous Vide Precision Cooker?

Take note that the head containing the display as well as the electronics should never be submerged in water, as this may result in damage to the device.

The user can clean their Anova precision cooker quite easily, and they can choose to clean specific portions as well. If they wish to clean the skirt and end cap of their Anova precision cooker, they have to disassemble the device, but the skirt and end cap themselves are dishwasher safe, and they can simply be placed in the dishwasher or hand washed in the sink.

If there is a build up on the device, this is most likely due to impurities in the water used to cook *sous vide*. This build up can take the form of brown, green, or black grime on the body of the Anova precision cooker, and this can easily be taken care of through a vinegar bath. A vinegar bath is done by mixing equal parts water and vinegar, and filling a pot with this solution. Clip the Anova Precision Cooker to the pot, and set it to work at a temperature of sixty Celsius / one hundred and forty Fahrenheit (60 C / 140 F). Once the mixture has reached that temperature, the device can be turned off and taken out, as the cleaning is complete. At this point, the buildup of minerals should have been removed.

If the user wishes to clean the other parts of the Anova Precision cooker, such as the circulator, sensors, heating coil, and other sensitive parts, the user has to disassemble the device. The user should take extra care to ensure that water does not reach these parts in order to avoid damage. In order to clean these specific parts, a damp cloth should suffice, and make sure that they are wiped dry immediately afterwards. In case a damp cloth is not enough, a mild soap and a soft brush may be used to clean. When cleaning the internal parts, take care in handling them, as some parts, such as the impeller shaft, may bend if handled roughly, preventing the device from functioning properly once reassembled.

For the Anova Precision Cooker's display, wiping it with a damp microfiber cloth should suffice to keep it clean. Once all the dirt or residue has been wiped off the display, make sure to dry the display gently with another microfiber cloth.

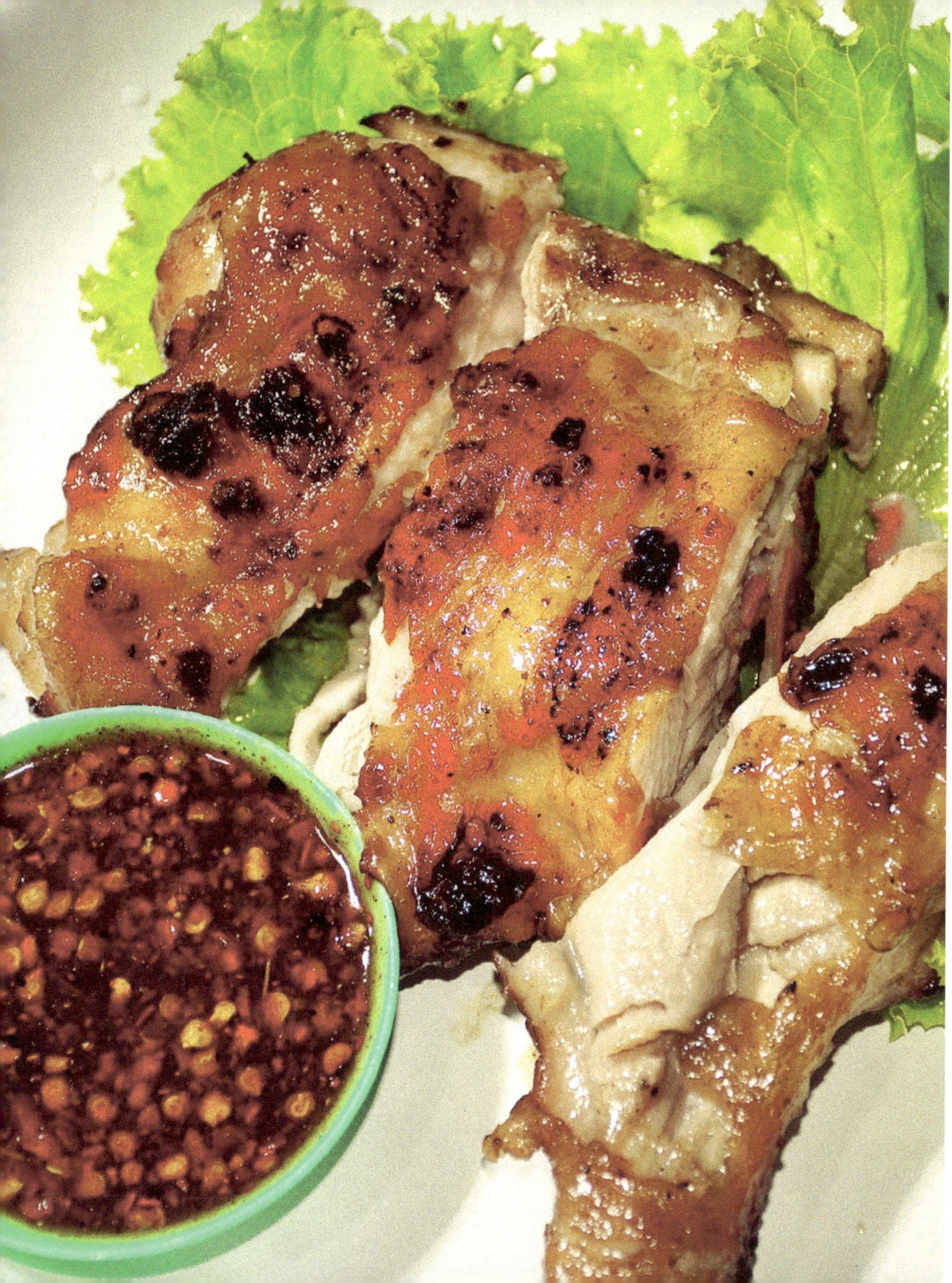

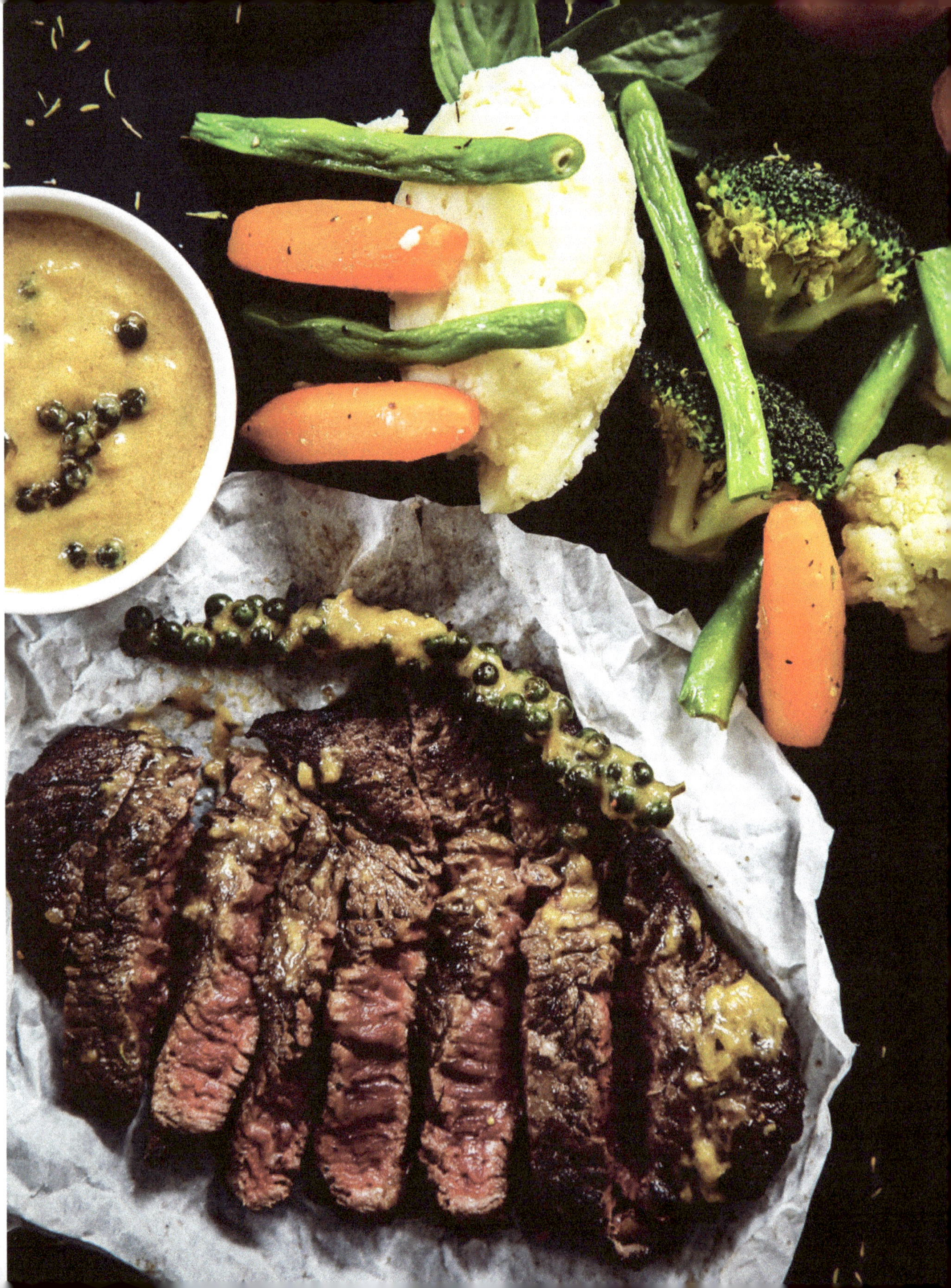

Chapter 1: Beef Recipes

Herb Butter Garlic Steaks 18

Beef Brisket 19

Rib Eye Steak 20

Beef Roast 21

Meatballs 22

Steak Strips with Herbs 23

Flank Steak 24

Filet Mignon with Chimichurri 25

Beef Tenderloin with Butter 26

Steak with Chimichurri Sauce 27

Simple Flank Steak 28

Beef Meatballs 29

Barbacoa Tacos 30

Tender and Juicy Beef Brisket 31

Flavorful Bavette Steak 32

Herb Butter Garlic Steaks

Serves: 4 / Preparation time: 10 minutes / cooking time: 1 hour 10 minutes / Temperature: 130 F / 54 C.

4 Filet Mignon Steaks

2 tsp fresh leaf parsley, chopped

2 tbsp olive oil

1 garlic clove, minced

2 tbsp butter

1/8 tsp garlic powder

Pepper

Kosher salt

- Preheat the sous vide water oven to 130 F / 54 C.
- Season steaks with pepper, salt, and garlic powder.
- Place steaks in a zip-lock bag remove the air from the bag before sealing.
- Place bag into the hot water bath and cook for 1 hour.
- Meanwhile, prepare garlic butter. Mix butter with minced garlic add a pinch of salt and parsley.
- Remove bag from the water bath and transfer steaks to a pan and sear over high heat with olive oil.
- Sear steaks 1 minute on each side.
- Transfer steaks into a serving dish and top with butter.
- Serve and enjoy.

Per Serving: Calories: 461; Total Fat: 28.8g; Saturated Fat: 10.7g; Protein: 48.2g; Carbs: 0.4g; Fiber: 0.1g; Sugar: 0g;

Beef Brisket

Serves: 8 / Preparation time: 10 minutes / Cooking time: 48 hours / Temperature: 133 F / 56 C

4 lbs beef brisket

1/2 tsp mustard powder

1 tsp onion powder

1 1/2 tsp garlic powder

3 tsp smoked paprika

1 tbsp liquid smoke

1 tbsp Worcestershire sauce

1 tsp pepper

2 tsp salt

- Fill and preheat sous vide water oven to 133 F/ 56 C.
- Rub the mixture of liquid smoke and Worcestershire sauce over brisket.
- In a small bowl, mix together smoked paprika, garlic, onion and mustard powder, pepper and salt.
- Rub bowl mixture over brisket. Place the brisket into the zip-lock bag and remove all the air from the bag before sealing.
- Place the bag into a water bath and cook for 48 hours.
- Remove the brisket from bag and grill for 10 minutes.
- Slice and serve.

Per Serving: Calories: 430; Total Fat: 14.3g; Saturated Fat: 5.4g; Protein: 69.1g; Carbs: 1.7g; Fiber: 0.5g; Sugar: 0.7g;

Rib Eye Steak

Serves: 8 / Preparation time: 10 minutes / Cooking time: 2 hours / Temperature: 129 F/ 53 C

18 oz boneless rib eye steak

1 tbsp butter, unsalted

2 tbsp olive oil

1 lemon zest

1 rosemary sprig

1 garlic cloves, smashed

2 thyme springs

Black pepper

Kosher salt

- Preheat the sous vide water oven to 129 F / 53 C.
- Season the steak with pepper and salt. Place steak into a zip-lock bag with thyme, rosemary, garlic, and lemon.
- Remove all the air from the bag before sealing.
- Place bag into a water bath and cook for 2 hours.
- Remove the steak from bag and pat dry with paper towels.
- Add olive oil to the pan and heat over high heat, add steak and sear both the side for 2 minutes.
- Slice the steak and season with salt.
- Serve and enjoy.

Per Serving: Calories: 217; Total Fat: 16.4g; Saturated Fat: 5.9g; Protein: 17.4g; Carbs: 0.1g; Fiber: 0g; Sugar: 0g;

Beef Roast

Serves: 8 / Preparation time: 10 minutes / Cooking time: 24 hours / Temperature: 136 F / 57 C

3 1/2 lbs beef roast

1/2 tsp onion powder

1 tsp rosemary, minced

1/2 tsp mustard powder

2 garlic cloves, minced

1/2 tbsp Worcestershire

1/2 tsp pepper

1 tsp smoked paprika

2 1/2 tsp salt

- Preheat the sous vide water oven to 136 F / 57 C.
- In a small bowl, mix together garlic, rosemary, smoked paprika, onion powder, mustard, pepper, and salt.
- Rub the roast in Worcestershire sauce and spread the spice over the roast.
- Place the roast into a zip-lock bag and remove all the air from the bag before sealing.
- Place the bag into a water bath and cook for 24 hours.
- Serve and enjoy.

Per Serving: Calories: 374; Total Fat: 12.5g; Saturated Fat: 4.7g; Protein: 60.4g; Carbs: 1g; Fiber: 0.3g; Sugar: 0.3g;

Meatballs

Serves: 4 / Preparation time: 10 minutes / Cooking time: 2 hours / Temperature: 140 F / 60 C

1 egg

16 oz ground beef

1/8 tsp nutmeg

1/8 tsp allspice

1/4 cup onion, minced

1/2 cup breadcrumbs

1 tbsp fresh parsley

Pepper

Salt

For Sauce:

2 cup beef broth

1 cup heavy cream

1 cup mushrooms, sliced

1 tbsp Dijon mustard

5 tbsp butter

5 tbsp flour

2 tbsp Worcestershire sauce

1 tbsp olive oil

- In a bowl, mix together ground beef, egg, onion and spices.
- Make small balls from mixture and place into refrigerator for 15 minutes.
- Place meatballs into a zip-lock bag and remove all the air from the bag before sealing.
- Place the bag into a water bath for 1 1/2 hours.
- Remove meatballs from bag and sear over medium-high heat with olive oil for 2 minutes or until lightly brown.
- Remove meatballs from pan and set aside. Melt butter in a pan stir in flour and whisk until creamy.
- Slowly stir in cream and broth and whisk in Dijon mustard and Worcestershire.
- Add mushrooms and cook for 2-3 minutes until mushrooms are soft.
- Now add meatballs and stir well.
- Serve and enjoy.

Per Serving: Calories: 613; Total Fat: 38.9g; Saturated Fat: 19.9g; Protein: 42.6g; Carbs: 21.6g; Fiber: 1.4g; Sugar: 3.5g;

Steak Strips with Herbs

Serves: 4 / Preparation time: 5 minutes / Cooking time: 2 hours / Temperature: 135 F / 57 C

2 steak strip, 1 inch thick

2 sprig thyme

2 sprig rosemary

2 tbsp butter

Black pepper

Salt

- Preheat water oven to 135 F / 57 C.
- Place the steak strips into the zip-lock bag with thyme, rosemary, and butter. Remove all the air from the bag before sealing.
- Place bag into a water bath and cook for 1 hour.
- Remove steak from bag and sear in a pan over high heat for 30 seconds on each side or until lightly brown.
- Serve and enjoy.

Per Serving: Calories: 130; Total Fat: 8.4g; Saturated Fat: 4.7g; Protein: 13g; Carbs: 0g; Fiber: 0g; Sugar: 0g;

Flank Steak

Serves: 3 / Preparation time: 10 minutes / Cooking time: 2 hours / Temperature: 132 F/ 56 C

16 oz flank steak

1/2 tbsp fish sauce

1 tbsp lemon juice

1/4 tsp cumin

2 tsp sugar

1 tsp salt

2 garlic cloves, chopped

2 tbsp soy sauce

2 tbsp olive oil

- In a bowl, mix together all ingredients except steak.
- Add steak to the marinade and mix well. Then Cover the bowl and refrigerate for 2 hours.
- Preheat water oven to 132 F / 56 C.
- Place steak into the zip-lock bag and remove all the air from the bag before sealing.
- Place bag in the hot water bath for 2 hours.
- Remove the steak from bag and sear in a pan from both the sides until lightly golden brown.
- Serve and enjoy.

Per Serving: Calories: 395; Total Fat: 22g; Saturated Fat: 6.6g; Protein: 43.1g; Carbs: 4.4g; Fiber: 0.2g; Sugar: 3.1g;

Filet Mignon with Chimichurri

Serves: 8 / Preparation time: 10 minutes / Cooking time: 1 hour 10 minutes / Temperature: 130 F / 54 C

- 28 oz filet mignon steaks cut 2 inches thick
- 2 tbsp butter, melted
- Black pepper
- 1 rosemary sprig
- 1 garlic clove, smashed
- 2 bacon strips
- Kosher salt
- For Chimichurri:
- 3 tsp sherry wine vinegar
- 3 tbsp lemon juice
- 3/4 cup grapeseed oil
- 2 tbsp green onion, minced
- 1 tsp ground pepper
- 2 tbsp fresh oregano leaves
- 1 tsp chili pepper flakes
- 1 tsp kosher salt
- 5 garlic cloves, minced
- 1/2 cup cilantro
- 1 cup parsley

- Set the water oven to 130 F / 54 C.
- Wrap steak with bacon strips and close with a toothpick. Season with black pepper and salt.
- Place steak in a zip-lock bag with rosemary, and garlic. Remove all the air from the bag before sealing.
- Place bag in the hot water bath for 1 hour.
- Remove the steak from bag and pat dry with paper towel.
- Sear steak with butter for 2 minutes in a pan.
- For chimichurri, add all ingredients into the food processor and process until combined. Serve over steak and enjoy.

Per Serving: Calories: 449; Total Fat: 35.3g; Saturated Fat: 8.2g; Protein: 30g; Carbs: 2.5g; Fiber: 1g; Sugar: 0.3g;

Beef Tenderloin with Butter

Serves: 6 / Preparation time: 10 minutes / Cooking time:12 hours / Temperature: 140 F / 60 C

24 oz center cut beef tenderloin

1 garlic clove, minced

2 tbsp fresh parsley leaves, chopped

7 tbsp butter

4 spring thyme

1 tbsp lemon juice

1 tbsp lemon zest

1 tbsp olive oil

- Cut tenderloin into four equal portions. Place cut side down and flatten gently.
- Season with Pepper and salt.
- Place the tenderloin pieces into a zip-lock bag with thyme spring. Remove all the air from the bag before sealing.
- Place bag into a water bath for 1 hour.
- Meanwhile, in a bowl, mix together 6 tbsp butter, parsley, garlic, lemon zest and lemon juice.
- Remove steak from the bag and pat dry with paper towel.
- Sear the steak with butter for 1 minute on each side in a pan until brown.
- Transfer the steak on a serving plate and top with a dollop of parsley butter.
- Serve and enjoy.

Per Serving: Calories: 375; Total Fat: 26.2g; Saturated Fat: 12.8g; Protein: 33.1g; Carbs: 0.5g; Fiber: 0.1g; Sugar: 0.1g;

Steak with Chimichurri Sauce

Serves: 6 / Preparation time: 10 minutes / Cooking time: 30 hours / Temperature: 134 F/ 56 C

2 lbs flank steak

Pepper

Salt

For chimichurri sauce:

2 cups parsley

1/3 tsp red pepper flakes

1 tbsp white vinegar

2 tbsp lemon juice

1/2 cup olive oil

1/4 medium onion, chopped

1 lemon zest

3 garlic cloves

1/2 cup mint leaves

- Fill and preheat sous vide water oven to 134 F/ 56 C.
- Season flank steak with pepper and salt.
- Place steak into the bag and remove all the air from the bag before sealing.
- Place bag into the hot water bath and cook for 30 hours.
- Meanwhile, for sauce add all sauce ingredients to the blender and blend until smooth.
- Remove steak from bag and cut into slices.
- Top steak slices with chimichurri sauce and enjoy.

Per Serving: Calories: 454; Total Fat: 29.7g; Saturated Fat: 7.7g; Protein: 43.1g; Carbs: 3g; Fiber: 1.4g; Sugar: 0.5g;

Simple Flank Steak

Serves: 4 / Preparation time: 10 minutes / Cooking time: 10 hours / Temperature: 140 F / 60 C

1 1/2 lbs flank steak

1 packet steak rub

Salt

- Fill and preheat sous vide water oven to 140 F/ 60 C.
- Season steak with salt and steak rub.
- Place steak into the zip-lock bag and remove all the air from the bag before sealing.
- Place bag into the hot water bath and cook for 10 hours.
- Remove meat from bag and pat dry with paper towel.
- Sear meat until lightly brown and serve.

Per Serving: Calories: 330; Total Fat: 14.2g; Saturated Fat: 5.9g; Protein: 47.3g; Carbs: 0g; Fiber: 0g; Sugar: 0g;

Beef Meatballs

Serves: 6 / Preparation time: 10 minutes / Cooking time: 3 hours / Temperature: 135 F / 57 C

1 lb ground beef

3 tbsp parmesan cheese, grated

1 tbsp garlic powder

1 tbsp dried oregano

3 tbsp parsley, chopped

1/2 shallot, diced

1 large egg, beaten

2 oz milk

1/4 cup breadcrumbs

1/4 tsp black pepper

1/2 tsp salt

- Fill and preheat sous vide water oven to 135 F/ 57 C.
- Add all ingredients to the mixing bowl and mix well to combine.
- Make small meatballs from the mixture.
- Place meatballs into the zip-lock bag and remove all the air from the bag before sealing.
- Place bag into the hot water bath and cook for 3 hours.
- Serve and enjoy.

Per Serving: Calories: 190; Total Fat: 6.8g; Saturated Fat: 2.4g; Protein: 25.7g; Carbs: 5g; Fiber: 0.7g; Sugar: 0.9g;

Barbacoa Tacos

Serves: 4 / Preparation time: 10 minutes / Cooking time: 8 hours / Temperature: 179 F / 82 C

1 lb beef chuck roast, trimmed and cut into pieces

1 tbsp chipotle in adobo, chopped

1 tbsp dried chilies

Black pepper

Salt

- Fill and preheat sous vide water oven to 179 F/ 82 C.
- Season beef with ground chilies, pepper, and salt.
- Transfer beef into the zip-lock bag and remove all the air from the bag before sealing.
- Place bag into the hot water bath and cook for 8 hours.
- Remove beef mixture from bag and place in the bowl.
- Using fork shred the meat and season with pepper and salt.
- Serve over tortillas and enjoy.

Per Serving: Calories: 413; Total Fat: 31.6g; Saturated Fat: 12.6g; Protein: 29.7g; Carbs: 0.3g; Fiber: 0g; Sugar: 0.1g;

Tender and Juicy Beef Brisket

Serves: 2 / Preparation time: 10 minutes / Cooking time: 24 hours / Temperature: 185 F / 85 C

1 lb beef brisket

1 bay leaf

2 garlic cloves, peeled

2 tbsp olive oil

4 thyme sprigs

1 celery stalk, chopped

1 onion, chopped

2 carrots, chopped

- Fill and preheat sous vide water oven to 185 F/ 85 C.
- Add brisket, bay leaf, garlic, butter, thyme, celery, onion, and carrot into the zip-lock bag and remove all the air from the bag before sealing.
- Place bag in the hot water bath for 24 hours.
- Remove meat from bag and using fork shred into bite-size pieces. Season with pepper and salt.
- Serve and enjoy.

Per Serving: Calories: 594; Total Fat: 28.2g; Saturated Fat: 7.3g; Protein: 70.2g; Carbs: 12.4g; Fiber: 2.9g; Sugar: 5.5g;

Flavorful Bavette Steak

Serves: 2 / Preparation time: 10 minutes / Cooking time: 2 hours 10 minutes / Temperature: 130 F / 54 C

16 oz bavette steak

4 garlic cloves, peeled

4 fresh thyme sprigs

4 tbsp butter

Black pepper

Salt

- Fill and preheat sous vide water oven to 130 F/ 54 C.
- Season steak with pepper and salt and place into the zip-lock bag.
- Remove all the air from the bag before sealing.
- Place bag in the hot water bath for 2 hours.
- Remove steak from bag and pat dry with paper towel.
- Heat a pan over medium-high heat.
- Sear steak in a pan until lightly golden brown.
- Turn the steak and add butter, garlic, and thyme to the pan. Cook steak until butter is melted.
- Serve and enjoy.

Per Serving: Calories: 664; Total Fat: 34.4g; Saturated Fat: 18.5g; Protein: 82.6g; Carbs: 2g; Fiber: 0.1g; Sugar: 0.1g;

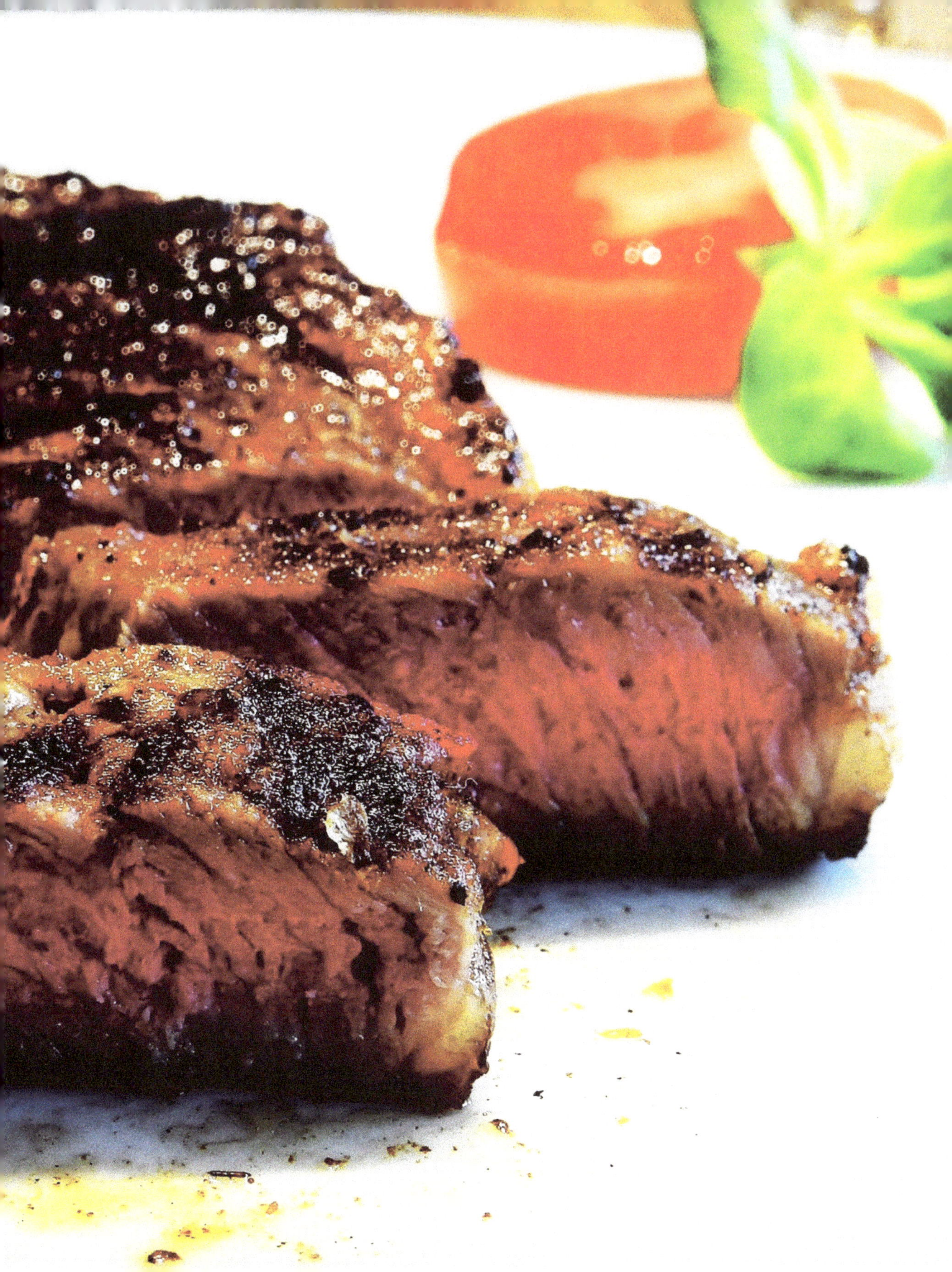

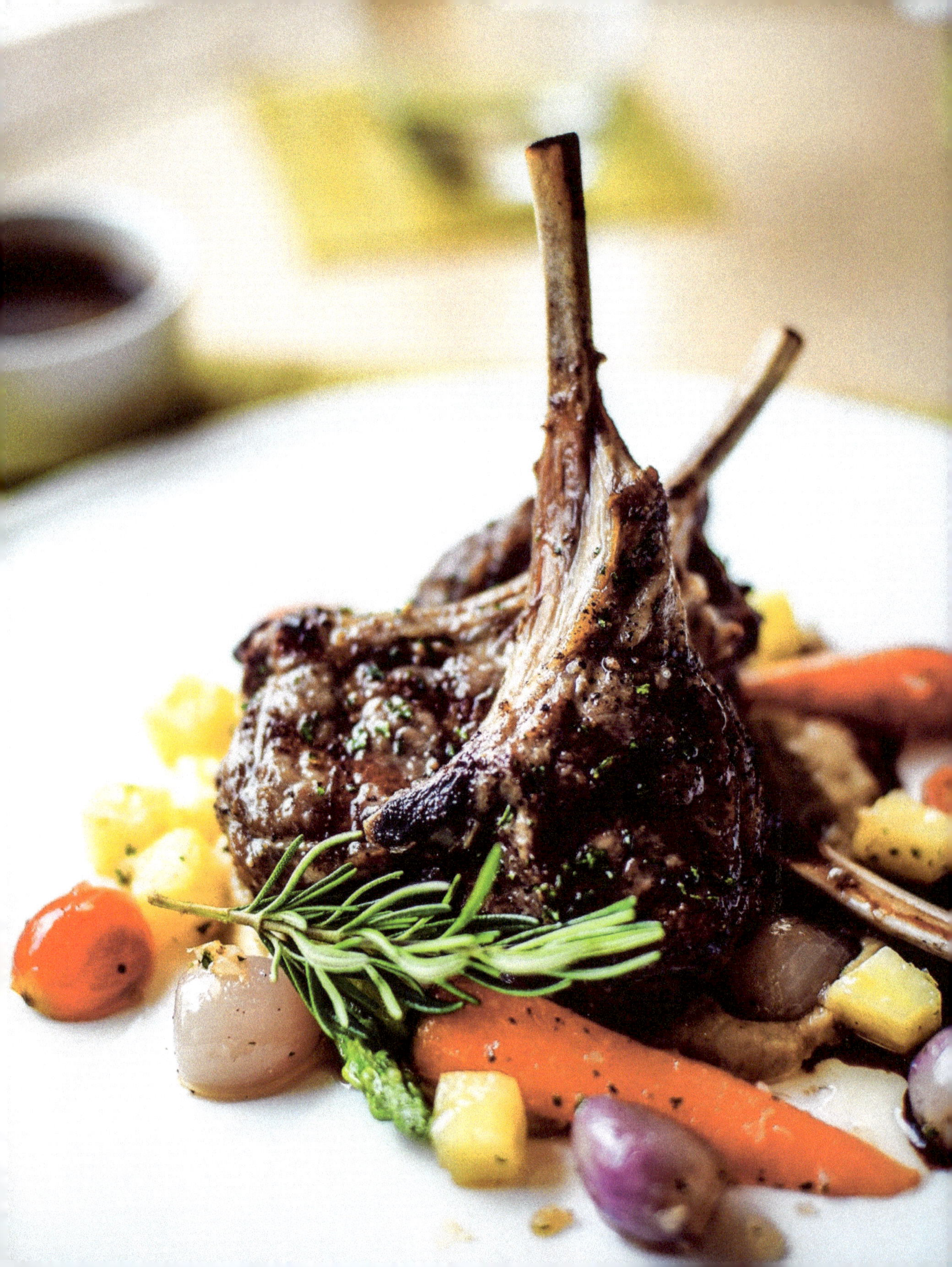

Chapter 2: Lamb Recipe

Rosemary Garlic Lamb Chops 35

Lamb Chops with Basil Chimichurri 37

Garlic Butter Lamb Chops 38

Simple Rack of Lamb 39

Herb Garlic Lamb Chops 40

Lamb Rack with Herb Butter 41

Lamb Burgers 42

Simple Spice Lamb 43

Lamb Loin with Mint Olive Salsa 44

Thyme Rosemary Lamb 45

Lamb with Mint Gremolata 46

Tender Lamb Chops 47

Soy Lemon Lamb Rack 48

Lamb Steaks 49

Meatballs with Sauce 50

Rosemary Garlic Lamb Chops

Serves: 4 / Preparation time: 5 minutes / Cooking time: 2 hours 30 minutes / Temperature: 140 F / 60 C

4 lamb chops

1 tbsp butter

1 tsp fresh thyme

1 tsp fresh rosemary

2 garlic cloves

Pepper

Salt

- Fill and preheat sous vide water oven to 140 F/ 60 C.
- Season lamb chops with pepper and salt.
- Sprinkle lamb chops with garlic, thyme, and rosemary.
- Add butter to the zip-lock bag then place lamb chops into the bag.
- Remove all air from the bag before sealing.
- Place bag in hot water bath and cook for 2 1/2 hours.
- Once it done then sear it on high heat until lightly brown.
- Serve and enjoy.

Per Serving: Calories: 349; Total Fat: 29g; Saturated Fat: 12.9g; Protein: 19.2g; Carbs: 0.9g; Fiber: 0.3g; Sugar: 0g;

Lamb Chops with Basil Chimichurri

Serves: 4 / Preparation time: 10 minutes / Cooking time: 2 hours / Temperature: 132 F / 56 C

2 rack of lamb, frenched
2 garlic cloves, crushed
Pepper
Salt
For basil chimichurri:
3 tbsp red wine vinegar
1/2 cup olive oil
1 tsp red chili flakes

2 garlic cloves, minced
1 shallot, diced
1 cup fresh basil, chopped
1/4 tsp pepper
1/4 tsp sea salt

- Fill and preheat sous vide water oven to 132 F/ 56 C.
- Season lamb with pepper and salt.
- Place lamb in a large zip-lock bag with garlic and remove all air from the bag before sealing.
- Place bag into the hot water bath and cook for 2 hours.
- Add all chimichurri ingredients to the bowl and mix well. Place in refrigerator for minutes.
- Remove lamb from bag and pat dry with paper towel.
- Sear lamb in hot oil. Sliced lamb between the bones.
- Place seared lamb chops on serving dish and top with chimichurri.
- Serve and enjoy.

Per Serving: Calories: 435; Total Fat: 44.8g; Saturated Fat: 12.1g; Protein: 8.4g; Carbs: 1.4g; Fiber: 0.2g; Sugar: 0.1g;

Garlic Butter Lamb Chops

Serves: 4 / Preparation time: 10 minutes / Cooking time: 2 hours / Temperature: 140 F / 60 C

4 lamb chops

1/2 tsp onion powder

1 garlic clove, minced

2 tbsp butter

1 tsp dried rosemary

Pepper

Salt

- Fill and preheat sous vide water oven to 140 F/ 60 C.
- Season chops with rosemary, pepper, and salt.
- Place lamb chops into the zip-lock bag and remove all air from the bag before sealing.
- Place bag in hot water bath and cook for 2 hours.
- Remove lamb chops from bag and pat dry with paper towel.
- In a microwave-safe bowl, add butter and microwave until butter is melted. Make sure butter doesn't burn it.
- Add onion powder and garlic into the melted butter and stir well.
- Baste lamb chops with butter mixture and sear until golden brown.
- Serve and enjoy.

Per Serving: Calories: 662; Total Fat: 29.8g; Saturated Fat: 12.2g; Protein: 92g; Carbs: 0.7g; Fiber: 0.2g; Sugar: 0.1g;

Simple Rack of Lamb

Serves: 4 / Preparation time: 5 minutes / Cooking time: 2 hours / Temperature: 140 F / 60 C

2 lbs rack of lamb

2 tbsp butter

2 tbsp canola oil

Black pepper

Salt

- Fill and preheat sous vide water oven to 140 F/ 60 C.
- Season lamb with pepper and salt and place in large zip-lock bag.
- Remove all air from the bag before sealing.
- Place bag in hot water bath and cook for 2 hours.
- Remove lamb from bag and pat dry with paper towels.
- Heat canola oil in a pan over medium heat.
- Spread butter over lamb and sear lamb in hot oil until lightly brown.
- Serve and enjoy.

Per Serving: Calories: 494; Total Fat: 32.8g; Saturated Fat: 11.2g; Protein: 46.2g; Carbs: 0g; Fiber: 0g; Sugar: 0g;

Herb Garlic Lamb Chops

Serves: 4 / Preparation time: 10 minutes / Cooking time: 2 hours / Temperature: 132 F / 56 C

4 lamb chops, bone in

2 tbsp butter

8 black peppercorns

1 tsp fresh oregano

1 tbsp fresh parsley

1 bay leaf

4 fresh thyme sprigs

2 tsp garlic, sliced

Sea salt

- Fill and preheat sous vide water oven to 132 F/ 56 C.
- Add lamb, butter, peppercorns, herbs, and garlic into the large zip-lock bag and remove all the air from the bag before sealing.
- Place bag into the hot water bath and cook for 2 hours.
- Remove lamb chops from bag and pat dry with paper towels.
- Heat pan over high heat and sear lamb chops for 30 seconds on each side.
- Serve and enjoy.

Per Serving: Calories: 663; Total Fat: 29.8g; Saturated Fat: 12.2g; Protein: 92.1g; Carbs: 0.9g; Fiber: 0.3g; Sugar: 0g;

Lamb Rack with Herb Butter

Serves: 4 / Preparation time: 10 minutes / Cooking time: 2 hours 5 minutes / Temperature: 134 F / 56 C

- 2 lamb racks, frenched
- Pepper
- Salt

For Herb Butter:

- 1 tbsp parmesan cheese, grated
- 1 tsp fresh mint, minced
- 1 tsp fresh chives, minced
- 1 tsp fresh rosemary, minced
- 1/2 tsp onion powder
- 1 garlic clove, minced
- 2 tbsp butter

- Fill and preheat sous vide water oven to 134 F/ 56 C.
- Season lamb with pepper and salt.
- Place lamb into the zip-lock bag and remove all air from the bag before sealing.
- Place bag in hot water bath and cook for 2 hours.
- Remove lamb from bag and pat dry with paper towels.
- Add all herbs to the butter and mix well.
- Coat the lamb with herb butter and broil for 3 minutes.
- Serve and enjoy.

Per Serving: Calories: 368; Total Fat: 18.4g; Saturated Fat: 8.3g; Protein: 46.8g; Carbs: 0.8g; Fiber: 0.2g; Sugar: 0.1g;

Lamb Burgers

Serves: 4 / Preparation time: 5 minutes / Cooking time: 2 hours / Temperature: 140 F / 60 C

1 lb ground lamb

1 tsp seasoning

Black pepper

Salt

- Fill and preheat sous vide water oven to 140 F/ 60 C.
- Add all ingredients to the bowl and mix well.
- Make four round patties from mixture and place into the large zip-lock bag.
- Place bag into the hot water bath and cook for 2 hours.
- Remove patties from the bag and pat dry with paper towels.
- Sear patties with olive oil until nicely golden brown.
- Serve and enjoy.

Per Serving: Calories: 211; Total Fat: 8.3g; Saturated Fat: 3g; Protein: 31.8g; Carbs: 0g; Fiber: 0g; Sugar: 0g;

Simple Spice Lamb

Serves: 3 / Preparation time: 10 minutes / Cooking time: 3 hours / Temperature: 112 F / 44 C

3 lamb loins, trimmed

2 shallot slices

1 garlic clove

1 tsp toasted spice blend

3 tbsp olive oil

Black pepper

Kosher salt

- Fill and preheat sous vide water oven to 112 F/ 44 C.
- Add all ingredients to the large zip-lock bag and shake well.
- Remove all air from the bag before sealing.
- Place bag into the hot water bath and cook for 3 hours.
- Remove lamb from bag and sear on all sides.
- Serve and enjoy.

Per Serving: Calories: 511; Total Fat: 32.9g; Saturated Fat: 9.2g; Protein: 51.4g; Carbs: 0.4g; Fiber: 0g; Sugar: 0g;

Lamb Loin with Mint Olive Salsa

Serves: 4 / Preparation time: 10 minutes / Cooking time: 2 hours 10 minutes / Temperature: 135 F / 57 C

1 1/2 lbs lamb loin

3 tbsp butter

2 rosemary sprigs

Pepper

Salt

For salsa:

1 garlic clove, grated

1 lemon zest

1/4 cup fresh mint, chopped

3 tbsp green olives, chopped

1 tbsp orange juice

1/4 cup olive oil

Pepper

Salt

- Fill and preheat sous vide water oven to 135 F/ 57 C.
- Season lamb with pepper and salt.
- Place lamb into the zip-lock bag with 2 tbsp butter and rosemary.
- Remove all air from the bag before sealing.
- Place bag into the hot water bath and cook for 2 hours.
- Meanwhile, add all salsa ingredients to the bowl and mix well. Set aside.
- Remove lamb from bag and pat dry with paper towels.
- Add remaining butter to the pan and sear lamb for 1 minute on each side.
- Place lamb on serving dish and top with salsa.
- Serve and enjoy.

Per Serving: Calories: 539; Total Fat: 38g; Saturated Fat: 13g; Protein: 45g; Carbs: 1.5g; Fiber: 0.6g; Sugar: 0.4g;

Thyme Rosemary Lamb

Serves: 2 / Preparation time: 10 minutes / Cooking time: 2 hours / Temperature: 140 F / 60 C

1 lb frenched rack of baby lamb

1 oz roasted garlic oil

1 tbsp fresh rosemary, minced

1 tbsp thyme, minced

1 tbsp ground black pepper

1 tbsp kosher salt

- Fill and preheat sous vide water oven to 140 F/ 60 C.
- Season lamb with rosemary, thyme, pepper, and salt.
- Place lamb into the large zip-lock bag and remove all air from the bag before sealing.
- Place bag into the hot water bath and cook for 2 hours.
- Remove lamb from bag and pat dry with paper towel.
- Add garlic oil to the pan and seared lamb until lightly brown.
- Cut into pieces and serve.

Per Serving: Calories: 840; Total Fat: 48g; Saturated Fat: 22g; Protein: 26g; Carbs: 4g; Fiber: 2.1g; Sugar: 0g;

Lamb with Mint Gremolata

Serves: 4 / Preparation time: 10 minutes / Cooking time: 2 hours / Temperature: 135 F / 57 C

1 rack of lamb

1 tbsp canola oil

1 rosemary sprig

Black pepper

Salt

For Mint Gremolata:

2 tbsp olive oil

1/4 tsp red pepper flakes

1 small lemon zest

2 garlic cloves

1 cup mint leaves

1/4 tsp kosher salt

- Fill and preheat sous vide water oven to 135 F/ 57 C.
- Season lamb with pepper and salt.
- Place lamb with rosemary in the large zip-lock bag and remove all air from the bag before sealing.
- Place bag into the hot water bath and cook for 2 hours.
- Meanwhile, prepare mint gremolata. Add all gremolata ingredients into the food processor and process until finely chopped.
- Once lamb did then remove from bag and pat dry with paper towel.
- Heat canola oil in a pan over high heat and sear lamb until browned.
- Serve with mint gremolata and enjoy.

Per Serving: Calories: 151; Total Fat: 13g; Saturated Fat: 2g; Protein: 6g; Carbs: 2.5g; Fiber: 1.6g; Sugar: 0g;

Tender Lamb Chops

Serves: 2 / Preparation time: 5 minutes / Cooking time: 2 hours / Temperature: 131 F / 55 C

2 lamb loin chops

2 garlic cloves, minced

2 rosemary sprigs

4 thyme sprigs

- Fill and preheat sous vide water oven to 131 F/ 55 C.
- Season lamb chops with pepper and salt.
- Place lamb chops into the zip-lock bag and top with rosemary, thyme, and garlic.
- Remove all air from the bag before sealing.
- Place bag into the hot water bath and cook for 2 hours.
- Remove lamb from bag and pat dry with paper towel.
- Using kitchen torch sear the exterior of lamb chops.
- Serve immediately and enjoy.

Per Serving: Calories: 613; Total Fat: 24g; Saturated Fat: 8g; Protein: 92g; Carbs: 1g; Fiber: 0.1g; Sugar: 0g;

Soy Lemon Lamb Rack

Serves: 8 / Preparation time: 10 minutes / Cooking time: 2 hours / Temperature: 129 F / 54 C

1 1/2 lbs rack of lamb, frenched

1 tsp lemon zest

2 tsp soy sauce

1 tbsp canola oil

- Fill and preheat sous vide water oven to 129 F/ 54 C.
- Brush lamb with soy sauce and place in zip-lock bag with lemon zest.
- Remove all air from the bag before sealing.
- Place bag in hot water bath and cook for 2 hours.
- Remove lamb from bag and pat dry with paper towel.
- Heat oil in a pan over high heat and sear lamb until lightly brown.
- Cut into segments and serve.

Per Serving: Calories: 332; Total Fat: 31g; Saturated Fat: 12g; Protein: 12g; Carbs: 0.2g; Fiber: 0g; Sugar: 0g;

Lamb Steaks

Serves: 2 / Preparation time: 10 minutes / Cooking time: 3 hours / Temperature: 134 F / 56 C

4 lamb T-bone chops

2 tbsp olive oil

2 fresh rosemary sprigs

Black pepper

Kosher salt

- Fill and preheat sous vide water oven to 134 F/ 56 C.
- Season chops with pepper and salt.
- Place lamb chops into the zip-lock bag with olive oil and rosemary.
- Remove all air from the bag before sealing.
- Place bag in hot water bath and cook for 3 hours.
- Remove lamb from bag and sear on a hot pan until browned.
- Serve and enjoy.

Per Serving: Calories: 420; Total Fat: 26g; Saturated Fat: 6g; Protein: 46g; Carbs: 0g; Fiber: 0g; Sugar: 0g;

Meatballs with Sauce

Serves: 4 / Preparation time: 10 minutes / Cooking time: 2 hours / Temperature: 134 F / 56 C

- 1 lb ground lamb
- 1/4 tsp cayenne pepper
- 1 tsp lemon juice
- 3 tbsp fresh mint, chopped
- 1/2 cup cucumber, diced
- 1 cup yogurt
- 1/4 tsp ground cinnamon
- 2 tsp ground coriander
- 2 garlic cloves, minced
- 1/4 cup pine nuts, toasted and chopped
- 1/4 cup onion, minced
- 1/4 cup parsley, chopped
- Salt

- Fill and preheat sous vide water oven to 134 F/ 56 C.
- Add lamb, cinnamon, coriander, salt, garlic, pine nuts, onion, and parsley into the bowl and mix well to combine.
- Make 20 meatballs from mixture and place into the zip-lock bag.
- Remove all the air from the bag before sealing.
- Place bag into the hot water bath and cook for 2 hours.
- Meanwhile, for sauce combine all remaining ingredients into the medium bowl.
- Serve meatball with sauce and enjoy.

Per Serving: Calories: 323; Total Fat: 15g; Saturated Fat: 4g; Protein: 37g; Carbs: 7.9g; Fiber: 1.1g; Sugar: 5.2g;

Anova Sous Vide Cookbook

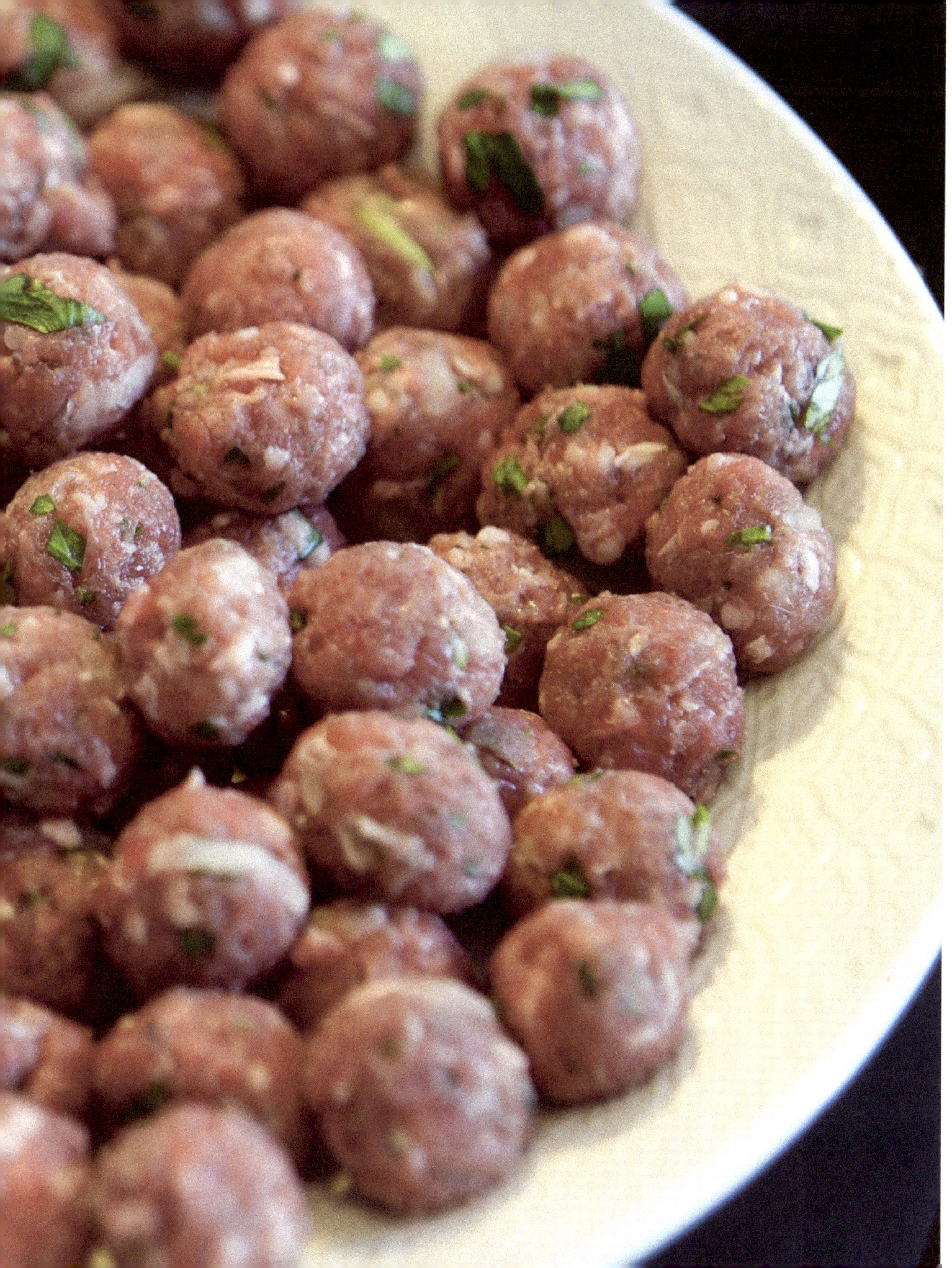

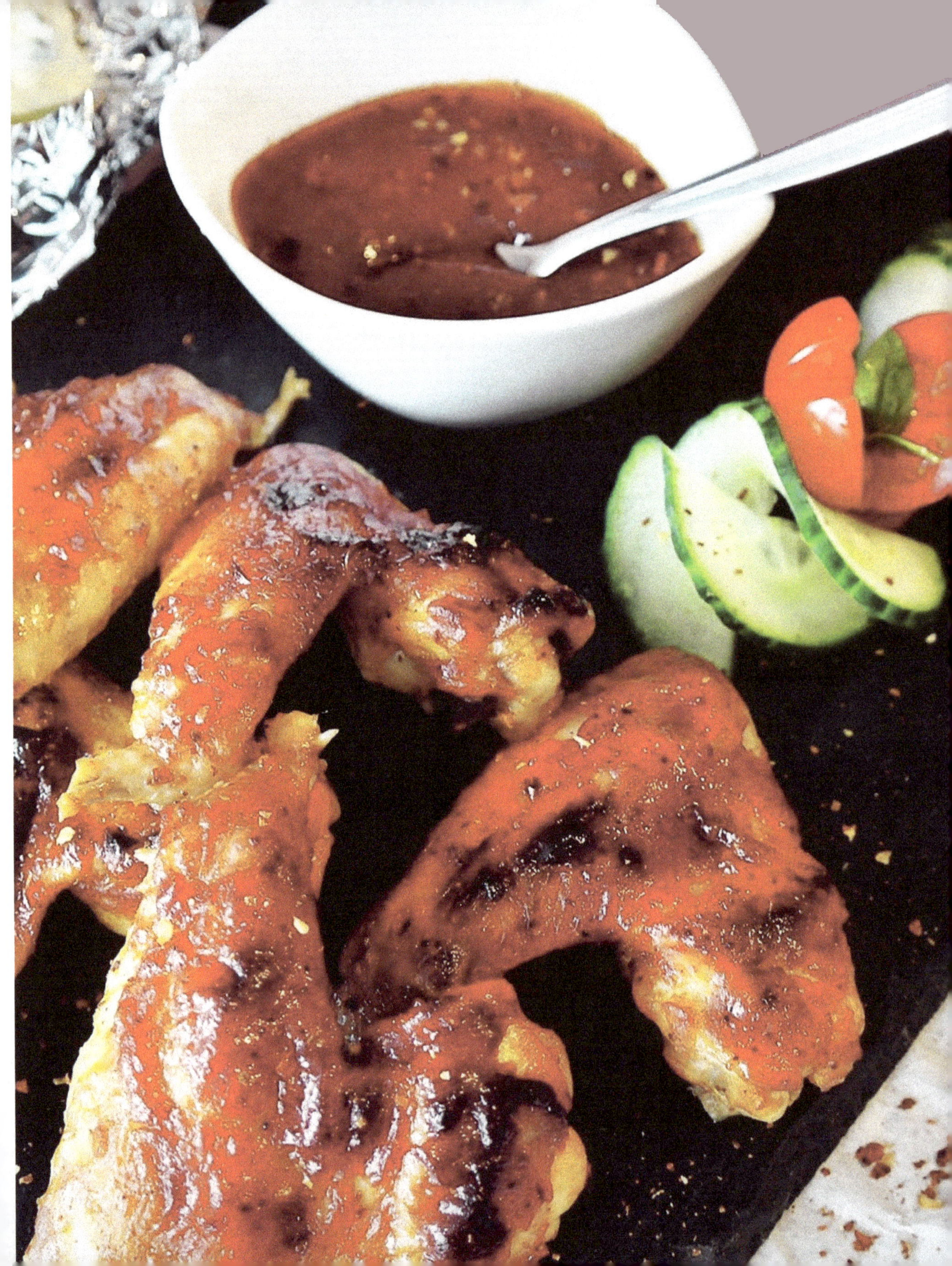

Chapter 3: Chicken Recipe

Simple Chicken Breast 54

Chicken Thighs 55

Chicken with Tomato Vinaigrette 56

Lemon Thyme Chicken 57

Balsamic Chicken Breast 58

Chicken Wings 59

Fried Chicken 60

Chicken Breast with Lemon Sauce 61

Chicken Fajitas 62

Chicken Ancho Chile 63

Chicken Meatballs 64

Chicken Adobo 65

Chicken Legs 66

Greek Chicken Meatballs 67

Simple No Sear Chicken Breast 68

Simple Chicken Breast

Serves: 2 / Preparation time: 5 minutes / Cooking time: 2 hours 10 minutes / Temperature: 140 F / 60 C

2 Chicken breast bone-in, skin-on

2 tbsp canola oil

Lemon

Pepper

Salt

- Season the chicken breast with pepper and salt.
- Add chicken breast in a zip-lock bag and remove all the air from the bag.
- Place bag into the hot water bath and cook for 2 hours.
- Remove chicken from bag and pat dry with paper towel.
- Heat oil in a pan over medium-high heat. Once the oil is hot then place chicken on pan skin side down.
- Cook until chicken until lightly brown and crisp.
- Remove chicken from pan and cut into slices.
- Serve the chicken with lemon slices.

Per Serving: Calories: 416; Total Fat: 20g; Saturated Fat: 2.8g; Protein: 55.3g; Carbs: 0.5g; Fiber: 0.1g; Sugar: 0.5g;

Chicken Thighs

Serves: 3 / Preparation time: 5 minutes / Cooking time: 2 hours 10 minutes / Temperature: 165 F / 74 C

4 chicken thighs

1 tbsp avocado oil

Freshly ground black pepper

Kosher salt

- Preheat the water bath to 165 F / 74 C.
- Season chicken thighs with salt and pepper.
- Place chicken thighs in single layer in a zip-lock bag and remove all the air from the bag before sealing.
- Place the bag into hot water bath for 2 hours.
- Heat oil in a pan over medium heat.
- Remove the chicken thighs from the bag and pat dry with paper towel.
- Sear chicken thighs until golden brown turn chicken thighs on the skin side and sear until crisp.
- Serve and enjoy.

Per Serving: Calories: 220; Total Fat: 13.9g; Saturated Fat: 4.1g; Protein: 25.4g; Carbs: 0.3g; Fiber: 0.2g; Sugar: 0g;

Chicken with Tomato Vinaigrette

Serves: 3 / Preparation time: 10 minutes / Cooking time: 2 hours 10 minutes / Temperature: 140 F / 60 C

4 chicken breasts, skin-on
Black pepper
Kosher salt
1 poblano pepper
1 tbsp canola oil
1 tbsp fresh mint leaves, minced
2 tsp lemon juice

1 medium shallot, minced
1/2 tsp hot sauce
1/2 tsp soy sauce
1 tsp honey
1/2 cup sun-dried tomatoes, drained and chopped

- Preheat water oven to 140 F / 60 C.
- Season chicken with black pepper and kosher salt.
- Place the chicken into zip-lock bag and remove all the air from the bag before sealing.
- Place bag into the hot water bath and cook for 2 hours.
- Meanwhile, set the burner to high heat place poblano pepper directly on the flame. Cook up to 5 minutes.
- Place poblano pepper in a paper bag until skin is loosened. Peel the skin under cool water, remove stem and seeds.
- Cut into 1/4 inch pieces and combine it with chopped sun-dried tomatoes, oil, hot sauce, soy sauce, honey lemon juice, shallot and mint leaves. Mix well.
- Add oil in a pan and heat over medium heat.
- Remove the chicken breast from the bag and pat dry with paper towel.
- Add chicken to a pan and cook for 2-3 minutes or until brown and crisp.
- Transfer chicken to a plate top with vinaigrette.
- Serve and enjoy.

Per Serving: Calories: 577; Total Fat: 25.1g; Saturated Fat: 6.1g; Protein: 78.6g; Carbs: 4.8g; Fiber: 0.8g; Sugar: 3.6g;

Lemon Thyme Chicken

Serves: 2 / Preparation time: 5 minutes / Cooking time: 2 hours / Temperature: 140 F / 60 C

2 chicken breasts

3 garlic cloves, chopped

7 spring of thyme leaves

1 1/2 tbsp olive oil

1 thinly sliced lemon

Pepper

Salt

- In a bowl, add olive oil, thyme, garlic, salt, and pepper with chicken breasts and mix well.
- Cover the bowl and place in refrigerator for 2 hours.
- When ready to cook place chicken breast in a zip-lock bag with lemon slices. Remove all the air from the bag before sealing.
- Place the bag into a water bath and cook for 2 hours.
- Remove the chicken from bag and sear in a pan on both the sides.
- Garnish with fresh thyme and lemon.
- Serve and enjoy.

Per Serving: Calories: 225; Total Fat: 13.3g; Saturated Fat: 1.5g; Protein: 24g; Carbs: 1.5g; Fiber: 0.1g; Sugar: 0.1g;

Balsamic Chicken Breast

Serves: 1 / Preparation time: 5 minutes / Cooking time: 1 hour 30 minutes / Temperature: 150 F / 65 C

1 chicken breast, large and boneless

1 orange, sliced

3 tbsp balsamic vinegar

1 rosemary sprig

Pepper

Salt

- Preheat the water oven 150 F / 65 C.
- Season the chicken with salt and pepper.
- Place chicken into the zip-lock bag with orange slices and rosemary. Remove all the air from the bag before sealing.
- Place bag in the hot water bath for 90 minutes.
- Serve hot and enjoy.

Per Serving: Calories: 224; Total Fat: 3g; Saturated Fat: 0g; Protein: 25.5g; Carbs: 22.1g; Fiber: 4.5g; Sugar: 17.4g;

Chicken Wings

Serves: 6 / Preparation time: 10 minutes / Cooking time: 2 hours / Temperature: 147 F / 64 C

24 chicken wings

1/2 cup sugar

1/2 cup fish sauce

1/4 cup warm water

1/4 cup tempura batter mix

1 cup rice flour

1-2 tsp chili garlic sauce

1 tsp kosher salt

10 garlic cloves, crushed

Canola oil

- Preheat water oven to 147 F / 64 C.
- In a bowl, mix together salt, warm water, fish sauce and sugar into crushed garlic.
- Place chicken into the zip-lock bag and remove all the air from the bag before sealing.
- Place bag in the hot water bath for 2 hours.
- Remove the wings from bag and pat dry with paper towel.
- Heat the oil to fry. Take a bowl and stir 1 cup rice flour and 1/4 cup tempura batter mix, dip the chicken wings in mixture
- Fry chicken wings for 3 minutes until lightly brown.
- Heat the pan and add remaining fish sauce plus sauce from the cooking bag, add chili garlic sauce on high heat reduce for 50 seconds
- Add chicken wings to sauce and serve.

Per Serving: Calories: 583; Total Fat: 26.9g; Saturated Fat: 7.5g; Protein: 40.1g; Carbs: 43.4g; Fiber: 0.7g; Sugar: 17.6g;

Fried Chicken

Serves: 6 / Preparation time: 10 minutes / Cooking time: 1 hour 10 minutes / Temperature: 165 F / 73 C

For Marinade and chicken

35 oz chicken drumstick

1 tsp five-spice powder

1 tsp garlic powder

1 tsp salt

1 tbsp oyster sauce

For Batter:

1/2 cup all-purpose flour

1/2 cup rice flour

1/2 cup cornstarch

1 cup cold water

1 tsp baking soda

1 tsp salt

- Preheat water oven to 165 F / 73 C.
- Mix all the marinade ingredients in the bowl.
- Add chicken to the marinade and place in the fridge for half hour.
- Place chicken into the zip-lock bag and remove all the air from the bag before sealing.
- Place the bag into the hot water bath for 1 hour.
- Remove chicken from bag and degrade with flour, whisk all batter ingredients in a bowl.
- Dip the chicken pieces in batter and deep fry until golden brown.
- Serve and enjoy

Per Serving: Calories: 408; Total Fat: 9.7g; Saturated Fat: 2.6g; Protein: 47.5g; Carbs: 28.7g; Fiber: 0.8g; Sugar: 0.2g;

Chicken Breast with Lemon Sauce

Serves: 4 / Preparation time: 10 minutes / Cooking time: 1 hour 30 minutes / Temperature: 140 F / 60 C

2 chicken breasts

1 tsp lemon

1/4 cup chicken stock

1/4 cup wine

1 lemon, sliced

1 tsp Mustard

Fresh Parsley, garnish

Pepper

Salt

- Preheat your water oven to 140°F / 60°C.
- Season chicken breast with salt and pepper.
- Place season chicken into zip-lock bag with lemon slices. Remove all the air from the bag before sealing.
- Place bag into the hot water bath and cook for 1 1/2 hours.
- Remove chicken from bag and sear chicken in butter.
- Once the chicken is nice sear take the breast out of the pan.
- Add wine to the pan and cook until reduce the half add chicken stock and mustard allow it to reduce 1/4.
- Slice chicken and serve with mashed potatoes.
- Serve and enjoy.

Per Serving: Calories: 156; Total Fat: 5.7g; Saturated Fat: 1.5g; Protein: 21.4g; Carbs: 0.9g; Fiber: 0.2g; Sugar: 0.3g;

Chicken Fajitas

Serves: 4 / Preparation time: 10 minutes / Cooking time: 4 hours / Temperature: 146 F / 63 C

32 oz chicken breast, thinly sliced

3/4 cup water

1 onion, sliced

2 green peppers, sliced

2 packet fajita seasonings

For toppings:

Cheddar cheese

Sour cream

Salsa

Tortilla shell

- Preheat the sous vide water oven to 146 F / 63 C.
- Place sliced chicken pieces into the zip-lock bag. In a separate bowl mix fajita seasoning and water.
- Place onion and pepper in each bag and pour half the fajita seasoning mixture in each bag.
- Remove all the air from the bag before sealing.
- Place bag into sous vide pot and cook for 1 hour.
- Remove bag from water and empty it into a pan and brown the chicken about 2-3 minutes on each side.
- Serve with tortilla shell and toppings.

Per Serving: Calories: 281; Total Fat: 5.8g; Saturated Fat: 0g; Protein: 48.9g; Carbs: 5.3g; Fiber: 1.6g; Sugar: 2.6g;

Chicken Ancho Chile

Serves: 6 / Preparation time: 10 minutes / Cooking time: 2 hours / Temperature: 146 F / 63 C

4 chicken breasts, boneless and skinless

2 tbsp oil

1 tsp salt

1 onion, peeled and quartered

3 Ancho chilies, dried

2 garlic cloves

- Preheat sous vide water oven to 146 F / 63 C.
- Soak the ancho chilies in hot water for 12-15 minutes. Remove the steam and discard.
- In a food processor puree the chilies with garlic, onion, oil, and salt until smooth paste forms.
- Rub ancho chili paste over chicken breast. Place the chicken breast into the zip-lock pouch.
- Seal the pouch and place into a water bath, cook for 2 hours.
- Serve and enjoy.

Per Serving: Calories: 234; Total Fat: 11.8g; Saturated Fat: 2.6g; Protein: 28.4g; Carbs: 2g; Fiber: 0.4g; Sugar: 0.8g;

Chicken Meatballs

Serves: 3 / Preparation time: 10 minutes / Cooking time: 2 hours 10 minutes / Temperature: 146 F / 63 C

16 oz ground chicken

1 tsp fresh oregano, minced

1 tsp kosher salt

1 tbsp olive oil

2 garlic cloves, minced

1/4 cup breadcrumbs

1/2 tsp lemon zest, grated

1/2 tsp black pepper, freshly ground

Lemon wedges for serving

- Preheat the sous vide water oven to 146 F / 63 C.
- In a bowl, combine chicken, garlic, olive oil, oregano, lemon zest, pepper, and salt. Add bread crumbs and mix well using your hand.
- Make 10-12 meatballs from the mixture. Place meatball into zip-lock bag carefully. Seal the bag.
- Place the bag in the water oven and cook for 2 hours.
- Transfer the meatballs to a baking sheet.
- Heat broiler to high and broil the meatballs 5-7 minutes until brown.
- Serve and enjoy.

Per Serving: Calories: 368; Total Fat: 16.4g; Saturated Fat: 3.9g; Protein: 45.2g; Carbs: 7.8g; Fiber: 0.8g; Sugar: 0.6g;

Chicken Adobo

Serves: 4 / Preparation time: 10 minutes / Cooking time: 4 hours 10 minutes / Temperature: 155 F / 68 C

24 oz chicken thighs and drumstick

1/4 cup vinegar

1/2 cup chicken broth

1 tbsp whole peppercorn

1 head garlic, crushed

1 cup soy sauce

6 pieces dried bay leaves

- Preheat the sous vide water oven to 155 F / 68 C.
- Place the chicken into zip-lock bag add garlic, peppercorn, soy sauce and bay leaves.
- Remove all air from the bag before sealing. Place bag in water bath and cook for 4 hours
- Remove the chicken from zip-lock bag and transfer to a cooking pan with half of marinade soy sauce mixture.
- Boil the liquid then add chicken broth and vinegar in pan, cover and cook for 6-7 minutes.
- Serve and enjoy.

Per Serving: Calories: 282; Total Fat: 15.2g; Saturated Fat: 4.6g; Protein: 33.1g; Carbs: 5.1g; Fiber: 0.5g; Sugar: 1.2g;

Chicken Legs

Serves: 4 / Preparation time: 10 minutes / Cooking time: 1 hour 50 minutes / Temperature: 145 F / 62 C

4 chicken legs

1 tsp mirin

1/4 cup teriyaki sauce

Black pepper

Salt

- Fill and preheat sous vide water oven to 145 F/ 62 C.
- Season chicken legs with pepper and salt.
- Add chicken, mirin, and teriyaki sauce into the zip-lock bag and remove all the air from the bag before sealing.
- Place bag in the hot water bath for 1 hour 45 minutes.
- Remove chicken from bag and place on a baking tray and broil for 3 minutes.
- Serve and enjoy.

Per Serving: Calories: 283; Total Fat: 15.3g; Saturated Fat: 4.2g; Protein: 30.7g; Carbs: 3.4g; Fiber: 0g; Sugar: 2.9g;

Greek Chicken Meatballs

Serves: 4 / Preparation time: 10 minutes / Cooking time: 2 hours 10 minutes / Temperature: 146 F / 63 C

- 1 lb ground chicken
- 1/4 cup breadcrumbs
- 1/2 tsp ground black pepper
- 1/2 tsp lemon zest, grated
- 1 tsp fresh oregano, minced
- 2 garlic cloves, minced
- 1 tbsp olive oil
- 1 tsp kosher salt

- Fill and preheat sous vide water oven to 146 F/ 63 C.
- In a bowl, combine together chicken, pepper, lemon zest, salt, oregano, garlic, and olive oil. Gently mix in breadcrumbs.
- Make small meatballs from mixture and place into the zip-lock bag.
- Remove all the air from the bag before sealing. Place bag into the hot water bath and cook for 2 hours.
- Remove meatballs from bag and place on a baking tray and broil for 5 minutes.
- Serve and enjoy.

Per Serving: Calories: 276; Total Fat: 12.3g; Saturated Fat: 2.9g; Protein: 33.9g; Carbs: 5.8g; Fiber: 0.6g; Sugar: 0.5g;

Simple No Sear Chicken Breast

Serves: 2 / Preparation time: 5 minutes / Cooking time: 2 hours / Temperature: 150 F / 68 C

2 chicken breasts, boneless and skinless

1 tsp garlic powder

Pepper

Salt

- Fill and preheat sous vide water oven to 150 F/ 68 C.
- Season chicken with garlic powder, pepper, and salt.
- Place chicken into the zip-lock bag and remove all the air from the bag before sealing.
- Place bag into the hot water bath and cook for 2 hours.
- Serve and enjoy.

Per Serving: Calories: 282; Total Fat: 10.8g; Saturated Fat: 3g; Protein: 42.5g; Carbs: 1.1g; Fiber: 0.2g; Sugar: 0.3g;

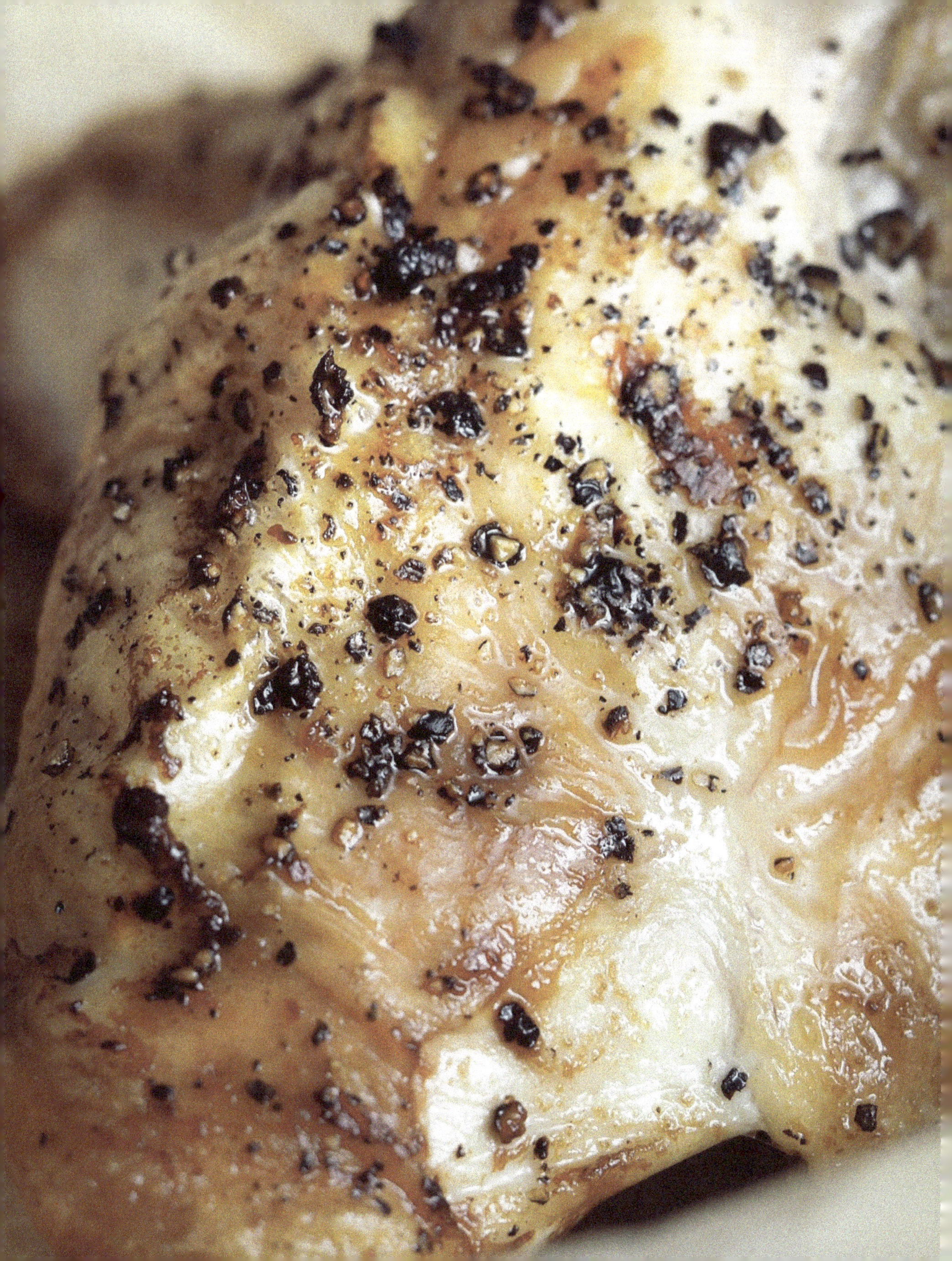

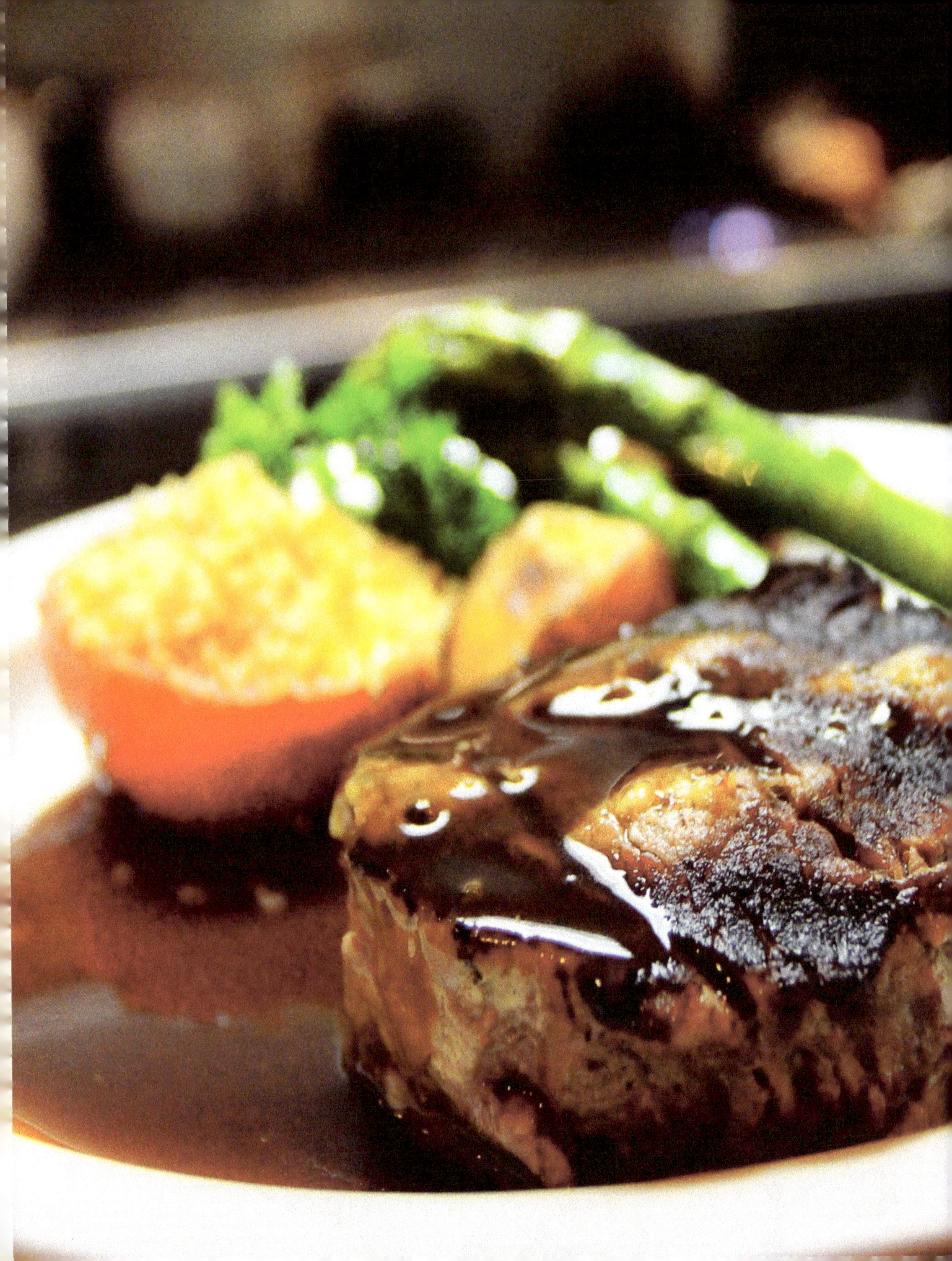

Chapter 4: Pork Recipes

Tender Pork Chops 72

Pork Chops with Mushrooms 73

Pork Tenderloin 74

Herb Rub Pork Chops 75

Bone- In Pork Chop 76

Pork Loin 77

Lemon Pork Chops 78

BBQ Pork Ribs 79

Five Spice Pork 80

BBQ Baby Back Ribs 81

Pork Carnitas 82

Pulled Pork 83

Simple Sliced Pork Belly 84

Perfect Pork Chop 85

Sweet and Spicy Pork Ribs 86

Tender Pork Chops

Serves: 2 / Preparation time: 5 minutes / Cooking time: 2 hours / Temperature: 140 F / 60 C

2 pork chops

1 tbsp canola oil

1 tbsp butter

4 garlic cloves

Rosemary

Thyme

Pepper

Salt

- Preheat water oven to 140 F / 60 C.
- Season the pork with pepper and salt. Place pork chops into the zip-lock bag.
- Remove all the air from the bag before sealing. Place bag into the hot water bath and cook for 2 hours.
- Remove pork from bag and pat dry with paper towel.
- Heat oil and butter in a pan over high heat with rosemary, thyme, and garlic.
- Place pork chops in a pan sear until lightly brown, about 1 minute on each side.
- Serve and enjoy.

Per Serving: Calories: 378; Total Fat: 32g; Saturated Fat: 11g; Protein: 18g; Carbs: 2g; Fiber: 0.1g; Sugar: 0.1g;

Pork Chops with Mushrooms

Serves: 4 / Preparation time: 10 minutes / Cooking time: 2 hours 10 minutes / Temperature: 140 F / 60 C

4 pork chops, boneless
4 tbsp butter
2 garlic cloves, minced
1 tbsp flour
1 cup chicken broth

8 oz crimini mushrooms, sliced
1 large shallot, sliced
Pepper
Salt

- Preheat water oven to 140 F / 60 C.
- Place the chops into a zip-lock bag. Place bag into a water bath and cook for 2 hours.
- Remove chops from bag and pat dry with paper towel.
- Season the chop with salt and pepper.
- Heat 2 tbsp butter in a pan. Sear the chops on both the sides.
- Add remaining 2 tbsp butter in a pan. Add sliced mushrooms to the pan and cook for 4-5 minutes stirring occasionally.
- Add shallots cook for 2 minutes until tender, add garlic and stir for 1 minute constantly then add flour.
- Stir well until mixture is evenly coated over mushrooms, add chicken broth stir for 1 minute.
- Season with salt and pepper.
- Serve and enjoy.

Per Serving: Calories: 392; Total Fat: 31.8g; Saturated Fat: 14.9g; Protein: 21g; Carbs: 4.6g; Fiber: 0.4g; Sugar: 1.2g;

Pork Tenderloin

Serves: 2 / Preparation time: 10 minutes / Cooking time: 2 hours 10 minutes / Temperature: 150 F / 66 C

16 oz pork tenderloin

1 tbsp olive oil

1 tbsp butter

2 small shallots, sliced

2 garlic cloves

8 sprigs fresh herbs

Black pepper

Kosher salt

- Preheat a water oven to 150 F / 66 C.
- Season the pork with salt and pepper.
- Place the pork into a zip-lock bag. Remove all the air from the bag before sealing.
- Place bag into the hot water bath and Cook for 2 hours.
- Remove pork from zip-lock bag and pat dry with paper towel.
- Heat olive oil in a pan over medium heat. Add pork and cook for 2 minutes until lightly browned.
- Add butter with fresh thyme, shallots, and garlic. cook for 1 minute.
- Serve and enjoy.

Per Serving: Calories: 440; Total Fat: 20.7g; Saturated Fat: 7.4g; Protein: 59.6g; Carbs: 1g; Fiber: 0.1g; Sugar: 0g;

Anova Sous Vide Cookbook

Herb Rub Pork Chops

Serves: 4 / Preparation time: 10 minutes / Cooking time: 2 hours 10 minutes / Temperature: 140 F / 60 C

4 pork chops, bone in

1/4 cup olive oil

1 tsp black pepper

1 tbsp balsamic vinegar

1 lemon zest

2 garlic cloves, minced

6 thyme sprigs, remove stems

1/4 cup chives

1/4 cup rosemary

10 basil leaves

1/4 cup parsley

1/2 tsp salt

- Fill and preheat sous vide water oven to 140 F/ 60 C.
- Add herbs to the food processor and process until chopped.
- Add garlic, olive oil, pepper, salt, vinegar, and lemon zest and blend until smooth paste.
- Rub herb mixture over pork chops. Place pork chops into the zip-lock bag and remove all the air from the bag before sealing.
- Place bag into the hot water bath and cook for 2 hours.
- Remove pork chops from the water bath and broil for 3-4 minutes.
- Serve and enjoy.

Per Serving: Calories: 383; Total Fat: 33.1g; Saturated Fat: 9.5g; Protein: 18.6g; Carbs: 3.6g; Fiber: 1.9g; Sugar: 0.1g;

Bone- In Pork Chop

Serves: 2 / Preparation time: 5 minutes / Cooking time: 2 hours 10 minutes / Temperature: 145 F / 62 C

2 pork chops, bone in

1 tsp olive oil

1/8 tsp tarragon

1/8 tsp thyme

Black pepper

Salt

- Fill and preheat sous vide water oven to 145 F/ 62 C.
- Season pork chops with pepper and salt.
- Rub tarragon, olive oil, and thyme over pork chops.
- Place pork chops into the zip-lock bag and remove all the air from the bag before sealing.
- Place bag into the hot water bath and cook for 2 hours.
- Remove pork chops from bag and sear until lightly brown.
- Serve and enjoy.

Per Serving: Calories: 276; Total Fat: 22.2g; Saturated Fat: 7.8g; Protein: 18g; Carbs: 0.1g; Fiber: 0g; Sugar: 0g;

Pork Loin

Serves: 4 / Preparation time: 5 minutes / Cooking time: 4 hours / Temperature: 153 F / 67 C

2 lbs pork loin roast

2 tbsp sweet and sour sauce

1 tsp black pepper

1 tsp garlic powder

1/2 tsp chipotle powder

1 tsp salt

- Fill and preheat sous vide water oven to 153 F/ 67 C.
- In a small bowl, mix together chipotle powder, garlic powder, black pepper, and salt.
- Rub spice mixture over the pork loin roast.
- Place pork into the zip-lock bag and remove all the air from the bag before sealing.
- Place bag into the hot water bath and cook for 4 hours.
- Remove pork from bag and coat outside with sweet and sour sauce.
- Broil pork for 5 minutes until lightly brown.
- Serve and enjoy.

Per Serving: Calories: 487; Total Fat: 21.9g; Saturated Fat: 8g; Protein: 65.1g; Carbs: 3.1g; Fiber: 0.3g; Sugar: 0.2g;

Lemon Pork Chops

Serves: 4 / Preparation time: 5 minutes / Cooking time: 6 hours / Temperature: 138 F / 59 C

4 pork chops, bone in

1 lemon, sliced

4 fresh thyme sprigs, chopped

1 tbsp olive oil

Pepper

Salt

- Fill and preheat sous vide water oven to 138 F/ 59 C.
- Season pork chops with pepper and salt.
- Place pork chops into the zip-lock bag with thyme and lemon slices. Drizzle with olive oil.
- Remove all air from the bag before sealing.
- Place bag into the hot water bath and cook for 6 hours.
- Remove pork chops from bag and pat dry with paper towel.
- Using kitchen torch sear the pork chops until caramelizing.
- Serve and enjoy.

Per Serving: Calories: 286; Total Fat: 23.4g; Saturated Fat: 8g; Protein: 18g; Carbs: 0g; Fiber: 0g; Sugar: 0g;

BBQ Pork Ribs

Serves: 2 / Preparation time: 5 minutes / Cooking time: 18 hours / Temperature: 160 F / 75 C

1 rack back ribs, cut into rib portions

2 tbsp Worcestershire sauce

1/3 cup brown sugar

1 1/2 cups BBQ sauce

- Fill and preheat sous vide water oven to 160 F/ 75 C.
- Whisk brown sugar in 1 cup BBQ sauce and Worcestershire sauce.
- Place ribs into the large mixing bowl then pour marinade over ribs and toss well.
- Place ribs into the zip-lock bag and remove all the air from the bag before sealing.
- Place bag into the hot water bath and cook for 18 hours.
- Remove ribs from bag and place on baking tray.
- Brush ribs with remaining BBQ sauce and broil for 5 minutes.
- Serve and enjoy.

Per Serving: Calories: 663; Total Fat: 19g; Saturated Fat: 6.5g; Protein: 10g; Carbs: 95.7g; Fiber: 1.1g; Sugar: 75.3g;

Five Spice Pork

Serves: 4 / Preparation time: 5 minutes / Cooking time: 48 hours / Temperature: 140 F / 60 C

1 lb pork belly

1 bacon slice

1 tsp Chinese 5 spice powder

Black pepper

Salt

- Fill and preheat sous vide water oven to 140 F/ 60 C.
- Add pork belly into the zip-lock bag with bacon slice and seasoning.
- Remove all the air from the bag before sealing.
- Place bag into the hot water bath and cook for 48 hours.
- Remove pork from bag and broil until crisp.
- Serve and enjoy.

Per Serving: Calories: 549; Total Fat: 32g; Saturated Fat: 13g; Protein: 54g; Carbs: 0.1g; Fiber: 0g; Sugar: 0g;

BBQ Baby Back Ribs

Serves: 2 / Preparation time: 5 minutes / Cooking time: 24 hours / Temperature: 143 F / 61 C

1 rack baby back pork ribs
6 tbsp Chipotle BBQ sauce
Pepper
Salt

- Fill and preheat sous vide water oven to 143 F/ 61 C.
- Cut rib rack in half and season with pepper and salt.
- Brush BBQ sauce over the pork ribs.
- Place ribs into the zip-lock bag and remove all the air from the bag before sealing.
- Place bag into the hot water bath and cook for 24 hours.
- Remove ribs from bag and grill for 1 minute.
- Serve and enjoy.

Per Serving: Calories: 850; Total Fat: 48g; Saturated Fat: 19g; Protein: 45g; Carbs: 59g; Fiber: 4.5g; Sugar: 10.5g;

Pork Carnitas

Serves: 12 / Preparation time: 10 minutes / Cooking time: 20 hours 10 minutes / Temperature: 175 F / 79 C

6 lbs pork shoulder

2 tbsp anise

2 bay leaves

2 cinnamon sticks

3 tbsp garlic, minced

4 bacon slices

1/3 cup brown sugar

2 orange juices

1 onion, chopped

1 tbsp sea salt

- Fill and preheat sous vide water oven to 175 F/ 79 C.
- In a small bowl, mix together anise, sugar, salt, garlic, and orange juice.
- Place pork into the zip-lock bag then pours orange juice mixture over pork.
- Add cinnamon, bay leaves, bacon, and onions into the bag.
- Seal bag and place into the hot water bath and cook for 20 hours.
- Heat large pan over medium-high heat.
- Remove pork from bag and place on pan and shred using a fork.
- Cook shredded pork until crispy.
- Serve and enjoy.

Per Serving: Calories: 729; Total Fat: 51g; Saturated Fat: 18g; Protein: 55g; Carbs: 7.6g; Fiber: 0.4g; Sugar: 5.5g;

Pulled Pork

Serves: 4 / Preparation time: 10 minutes / Cooking time: 18 hours 35 minutes / Temperature: 165 F / 73 C

2 lbs pork shoulder, boneless

1/2 cup taco seasoning

1 onion, diced

1/4 cup cilantro, chopped

- Fill and preheat sous vide water oven to 165 F/ 73 C.
- Season pork with half taco seasoning.
- Place pork into the zip-lock bag and remove all the air from the bag before sealing.
- Place bag into the hot water bath and cook for 18 hours.
- Remove pork from bag and pat dry with paper towel.
- Season pork with remaining taco seasoning.
- Place pork in preheated 350 F/ 176 C oven and cook for 30 minutes.
- Remove pork from oven and using fork shred the pork.
- Garnish with cilantro and serve.

Per Serving: Calories: 733; Total Fat: 48g; Saturated Fat: 17g; Protein: 53g; Carbs: 14g; Fiber: 0.6g; Sugar: 4.2g;

Simple Sliced Pork Belly

Serves: 2 / Preparation time: 10 minutes / Cooking time: 3 hours 10 minutes / Temperature: 145 F / 62 C

4 oz pork belly, sliced

3 bay leaves

1 tbsp garlic salt

1 tbsp whole black peppercorns

1 1/2 tbsp olive oil

- Fill and preheat sous vide water oven to 145 F/ 62 C.
- Add sliced pork belly, bay leaves, garlic salt, peppercorns, and 1 tbsp olive oil into the large zip-lock bag.
- Remove all the air from the bag before sealing.
- Place bag into the hot water bath and cook for 3 hours.
- Heat remaining oil in a pan over medium heat.
- Remove pork from bag and sear in hot oil for 2 minutes on each side.
- Serve and enjoy.

Per Serving: Calories: 374; Total Fat: 25g; Saturated Fat: 8g; Protein: 27g; Carbs: 5.1g; Fiber: 1.3g; Sugar: 1g;

Perfect Pork Chop

Serves: 2 / Preparation time: 5 minutes / Cooking time: 50 minutes / Temperature: 140 F / 60 C

20 oz pork rib chop, bone in

2 tbsp butter

Black pepper

Salt

- Fill and preheat sous vide water oven to 140 F/ 60 C.
- Season pork chops with pepper and salt.
- Place pork chops into the zip-lock bag and remove all the air from the bag before sealing.
- Place bag into the hot water bath and cook for 45 minutes.
- Heat butter into the pan over medium heat.
- Remove pork chop from the bag and pat dry with paper towel.
- Sear pork chops in hot butter until lightly brown from both the sides.
- Serve and enjoy.

Per Serving: Calories: 731; Total Fat: 48g; Saturated Fat: 20g; Protein: 69g; Carbs: 0g; Fiber: 0g; Sugar: 0g;

Sweet and Spicy Pork Ribs

Serves: 6 / Preparation time: 5 minutes / Cooking time: 20 hours 10 minutes / Temperature: 145 F / 62 C

2 full racks baby back pork ribs, cut in half

1/2 cup jerk seasoning mix

- Fill and preheat sous vide water oven to 145 F/ 62 C.
- Season pork rib rack with half jerk seasoning and place in large zip-lock bag.
- Remove all the air from the bag before sealing.
- Place bag into the hot water bath and cook for 20 hours.
- Remove meat from bag and rub with remaining seasoning and place on a baking tray.
- Broil for 5 minutes. Slice and serve.

Per Serving: Calories: 880; Total Fat: 56g; Saturated Fat: 21g; Protein: 56g; Carbs: 38g; Fiber: 3.3g; Sugar: 0g;

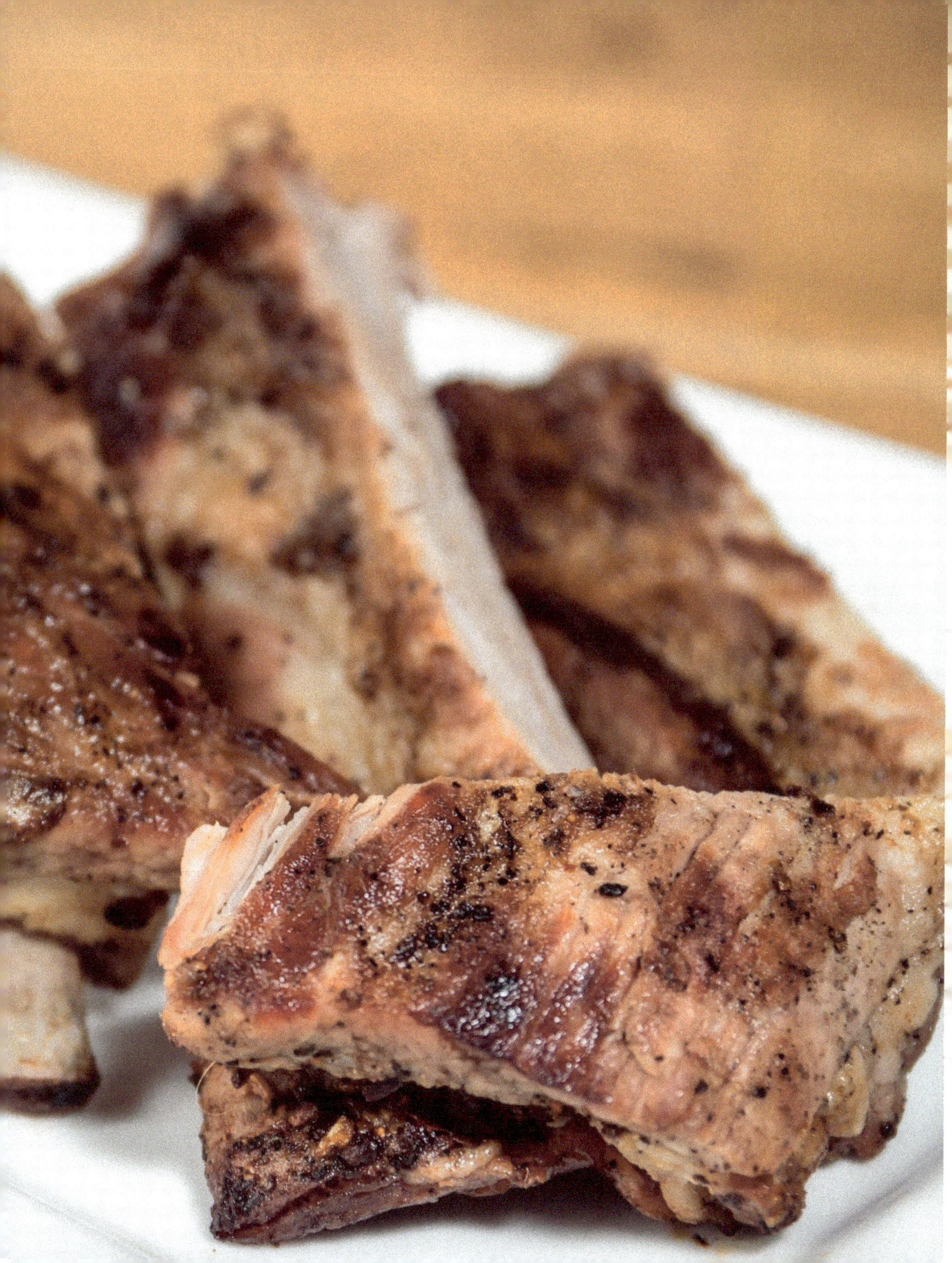

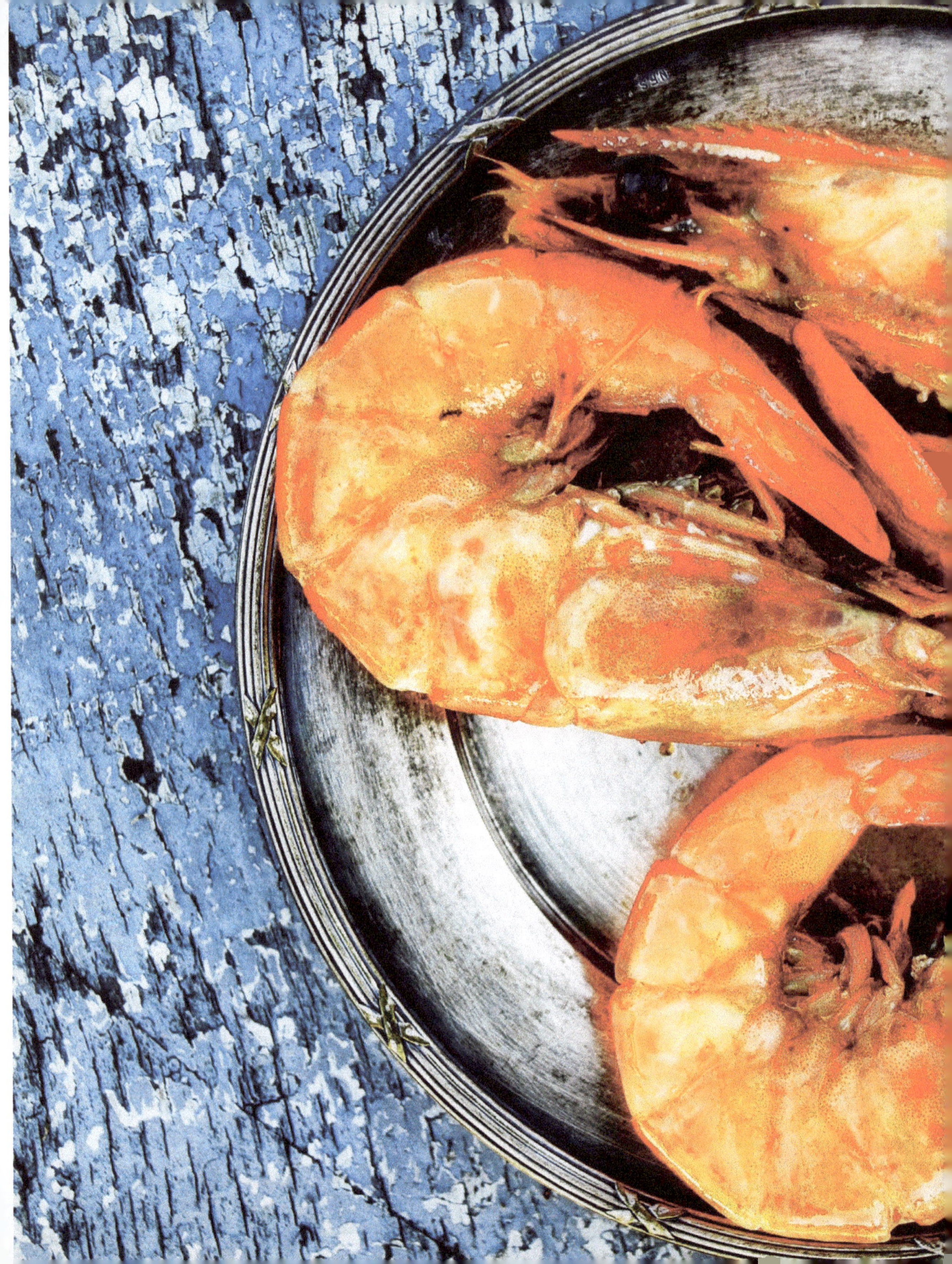

Chapter 5: Seafood Recipes

Smoked Paprika Shrimp 89

Scallops with Salsa Verde 91

Easy Pecan Salmon 92

Tuna 93

Buttery Halibut 94

Simple Brown Butter Scallops 95

BBQ Cranberry Salmon 96

Sous Vide Cod 97

Crispy Skin Salmon 98

Tasty Teriyaki Salmon 99

Juicy and Tender Swordfish 100

Bacon Wrapped Scallops 101

Honey Ginger Salmon 102

Poached Lobster 103

Mahi Mahi with Bean Puree 104

Smoked Paprika Shrimp

Serves: 6 / Preparation time: 10 minutes / Cooking time: 25 minutes / Temperature: 135 F / 57 C

1 1/2 lbs large shrimp, peeled
2 tbsp butter
1 1/2 tsp vinegar
3 tbsp sherry
2 bay leaves

1 tbsp smoked paprika
6 garlic cloves, sliced
6 tbsp olive oil
1/2 tsp baking soda
Salt

- Fill and preheat sous vide water oven to 135 F/ 57 C.
- In a mixing bowl, toss shrimp with baking soda and 1/2 tsp salt. Set aside.
- Heat olive oil in the pan over medium-low heat.
- Add garlic to the pan and sauté for 3 minutes.
- Add bay leaves and paprika and sauté for 30 seconds.
- Add vinegar and sherry and cook on high for 2 minutes or until liquid reduced.
- Remove pan from heat and stir in butter.
- Season with salt and set aside to cool for 5 minutes.
- Add shrimp in the large zip-lock bag then pour garlic and oil mixture into the bag.
- Remove all air from the bag before sealing.
- Shake bag well and place in preheated water bath and cook for 20 minutes.
- Remove bag from the water bath and pour shrimp in a large bowl.
- Serve and enjoy.

Per Serving: Calories: 253; Total Fat: 18g; Saturated Fat: 4g; Protein: 21g; Carbs: 3.7g; Fiber: 0.5g; Sugar: 0.2g;

Scallops with Salsa Verde

Serves: 4 / Preparation time: 10 minutes / Cooking time: 35 minutes / Temperature: 125 F/ 51 C

8 sea scallops, side muscle removed
1 garlic clove, minced
1 small shallot, minced
5 tbsp olive oil
1/2 lemon juice
1 tbsp cilantro, chopped
1 tbsp parsley, chopped

1 tbsp chives, chopped
1 tbsp basil, chopped
1 tbsp butter
Pepper
Salt

- Fill and preheat sous vide water oven to 125 F/ 51 C.
- Season scallops with pepper and salt.
- Add 1 tbsp olive oil and seasoned scallops into the large zip-lock bag.
- Remove all air from the bag before sealing.
- Shake bag once and place in water bath and cook for 30 minutes.
- Meanwhile, melt butter in a pan over medium heat.
- For salsa, add chopped herbs, lemon juice, garlic, and shallot into the bowl and mix well. Set aside for 5 minutes.
- Pour remaining olive oil over salsa. Mix well and set aside.
- Remove scallops from water bath and place on a plate. Pat dry with paper towel.
- Add scallops in butter pan and cook for 50 seconds on each side.
- Serve scallops with salsa and enjoy.

Per Serving: Calories: 230; Total Fat: 20g; Saturated Fat: 4g; Protein: 10g; Carbs: 1.8g; Fiber: 0.1g; Sugar: 0g;

Easy Pecan Salmon

Serves: 2 / Preparation time: 10 minutes / Cooking time: 35 minutes / Temperature: 125 F / 51 C

2 large salmon fillets

1/4 cup pecans cut into pieces

1/4 cup orange juice

1/4 cup bourbon

1/2 cup maple syrup

Pepper

Salt

- Fill and preheat sous vide water oven to 125 F/ 51 C.
- Place salmon fillets into the large zip-lock bag and remove all air from the bag before sealing.
- Place bag in hot water bath and cook for 30 minutes.
- Meanwhile, for sauce add all remaining ingredients into the saucepan and heat over medium heat and cook until sauce thickened.
- Once salmon is done then remove from water bath and place on serving the dish.
- Top with pecan sauce and serve.

Per Serving: Calories: 532; Total Fat: 12g; Saturated Fat: 1.7g; Protein: 34g; Carbs: 56g; Fiber: 0.3g; Sugar: 49g;

Tuna

Serves: 4 / Preparation time: 10 minutes / Cooking time: 35 minutes / Temperature: 115 F / 46 C

2 tuna steaks

2 tsp vegetable oil

1/2 cup sesame seeds

2 tbsp olive oil

Black pepper

Salt

- Fill and preheat sous vide water oven to 115 F/ 46 C.
- Season tuna with pepper and salt.
- Add tuna and oil into the zip-lock bag and remove all air before bag sealing.
- Place bag into the hot water bath and cook for 30 minutes.
- Heat vegetable oil into the pan over medium heat.
- Place sesame seeds in a dish and coat tuna with sesame seeds.
- Place tuna in a pan and cook for 45 seconds from each side.
- Serve and enjoy.

Per Serving: Calories: 265; Total Fat: 18g; Saturated Fat: 2.9g; Protein: 22g; Carbs: 4.7g; Fiber: 2.1g; Sugar: 0.2g;

Buttery Halibut

Serves: 4 / Preparation time: 10 minutes / Cooking time: 35 minutes / Temperature: 130 F / 54 C

1 1/2 lbs halibut fillets

4 tbsp butter

Black pepper

Salt

- Fill and preheat sous vide water oven to 130 F/ 54 C.
- Season halibut with pepper and salt.
- Place seasoned halibut and 2 tablespoons butter into the large zip-lock bag.
- Remove all air from the bag before sealing.
- Place bag into the hot water bath and cook for 35 minutes.
- Melt remaining butter into the pan over medium heat.
- Place halibut in a pan and cook for 40-45 seconds or until lightly browned.
- Serve and enjoy.

Per Serving: Calories: 291; Total Fat: 15g; Saturated Fat: 7.8g; Protein: 35.9g; Carbs: 0g; Fiber: 0g; Sugar: 0g;

Simple Brown Butter Scallops

Serves: 1 / Preparation time: 5 minutes / Cooking time: 35 minutes / Temperature: 140 F / 60 C

4.25 oz scallops

2 tsp brown butter

Pepper

Salt

- Fill and preheat sous vide water oven to 140 F/ 60 C.
- Season scallops with pepper and salt.
- Add scallops and 1 tsp brown butter into the zip-lock bag.
- Shake bag well and remove all air from the bag before sealing.
- Place zip-lock bag into the water bath and cook for 35 minutes.
- Heat remaining butter in the pan over medium heat.
- Seared scallops to the pan until lightly golden brown.
- Serve and enjoy.

Per Serving: Calories: 106; Total Fat: 0.9g; Saturated Fat: 0.1g; Protein: 20.2g; Carbs: 2.9g; Fiber: 0g; Sugar: 0g;

BBQ Cranberry Salmon

Serves: 2 / Preparation time: 10 minutes / Cooking time: 25 minutes / Temperature: 140 F / 60 C

10 oz salmon fillets, boneless

1 tsp lime juice

1 tbsp cranberry juice

1 tbsp olive oil

2 tbsp BBQ sauce

2 tbsp cranberry sauce

1/8 tsp salt

- Add all ingredients except salmon into the mixing bowl and mix well.
- Add salmon fillets to the marinade and place in refrigerator for 1 hour.
- Fill and preheat sous vide water oven to 140 F/ 60 C.
- Place salmon fillets into the large zip-lock bag.
- Remove all air from the bag before sealing.
- Place bag in hot water bath and cook for 25 minutes.
- Remove salmon from the water bath and broil in a preheated broiler for 2 minutes.
- Serve hot and enjoy.

Per Serving: Calories: 299; Total Fat: 15g; Saturated Fat: 2.3g; Protein: 27.5g; Carbs: 12.7g; Fiber: 0.4g; Sugar: 10.8g;

Sous Vide Cod

Serves: 2 / Preparation time: 10 minutes / Cooking time: 30 minutes / Temperature: 106 F / 41 C

2 cod steaks
4 tbsp butter
4 tbsp clarified butter
1 shallot, chopped
1/2 tsp sugar
1 cup dry white wine
1 tsp salt

- In a small bowl, mix together salt and sugar.
- Rub salt and sugar mixture over cod and place cod in the refrigerator for 1 hour.
- Heat clarified butter in a saucepan over medium heat.
- Add shallot to the pan and sauté for 15 minutes or until golden brown.
- Add white wine to the pan and cook until half wine is left.
- Remove cod from the refrigerator and wash salt and sugar. Pat dry cod with a paper towel.
- Fill and preheat sous vide water oven to 106 F/ 41 C.
- Add cod and shallot mixture to the large zip-lock bag.
- Remove all air from the bag before sealing.
- Place bag in hot water bath and cook for 30 minutes.
- Remove cod from zip-lock bag and set aside.
- Pour zip-lock bag sauce into the blender and blend until smooth.
- Pour sauce mixture into the saucepan and heat over medium-high heat until sauce thickens.
- Heat remaining butter in the pan over medium heat.
- Place cod in a pan and sear for 2 minutes.
- Place seared cod on serving dish and top with sauce.
- Serve and enjoy.

Per Serving: Calories: 603; Total Fat: 46g; Saturated Fat: 29g; Protein: 21g; Carbs: 4.3g; Fiber: 0g; Sugar: 2g;

Crispy Skin Salmon

Serves: 4 / Preparation time: 5 minutes / Cooking time: 15 minutes / Temperature: 125 F / 52 C

1 lb salmon, cut into pieces

1 tbsp ghee

2 tbsp olive oil

2 cups ice water

1/2 cup boiling water

1 tbsp honey

1/4 cup sea salt

- Add honey and salt to the boiling water and stir until dissolved. Add the ice water.
- Place salmon in a dish then pour brine over salmon. Refrigerate for 1 hour.
- After 1 hour remove salmon from brine and wash well. Pat dry with paper towel.
- Fill and preheat sous vide water oven to 125 F/ 52 C.
- Add salmon and olive oil into the zip-lock bag and remove all air from the bag before sealing.
- Place bag in hot water bath and cook for 15 minutes.
- Once done then remove from bag and pat dry with paper towel.
- Heat ghee in a pan over medium-high heat.
- Place salmon in a pan and sear for 2-3 minutes or until crispy.
- Serve and enjoy.

Per Serving: Calories: 254; Total Fat: 17g; Saturated Fat: 4g; Protein: 22g; Carbs: 4.3g; Fiber: 0g; Sugar: 4.3g;

Tasty Teriyaki Salmon

Serves: 4 / Preparation time: 10 minutes / Cooking time: 45 minutes / Temperature: 122 F / 50 C

1 1/2 lbs salmon fillets

2 tsp cornstarch

2 tbsp rice vinegar

2 tbsp brown sugar

1/4 cup water

1/4 cup soy sauce

4 tsp ginger, minced

3 garlic cloves, minced

- In a bowl, mix together vinegar, brown sugar, water, soy sauce, garlic, and ginger.
- Place salmon into the large zip-lock bag then pour sauce mixture into the bag.
- Remove all air from the bag before sealing.
- Fill and preheat sous vide water oven to 122 F/ 50 C.
- Place bag in hot water bath and cook for 40 minutes.
- Remove salmon from bag and place on a serving dish.
- Pour sauce into a saucepan and stir in cornstarch and cook over medium heat until thickened.
- Pour sauce over salmon and serve.

Per Serving: Calories: 270; Total Fat: 10g; Saturated Fat: 1.5g; Protein: 34g; Carbs: 8.8g; Fiber: 0.4g; Sugar: 4.7g;

Juicy and Tender Swordfish

Serves: 2 / Preparation time: 5 minutes / Cooking time: 35 minutes / Temperature: 130 F / 55 C

12 oz swordfish steaks

4 sprigs fresh thyme

2 lemons juice

2 tbsp olive oil

2 lemon zest

Black pepper

Kosher salt

- Fill and preheat sous vide water oven to 130 F/ 55 C.
- Season swordfish with pepper and salt.
- Place swordfish into the large zip-lock bag with thyme, lemon juice, lemon zest, and olive oil.
- Remove all air from the bag before sealing.
- Place bag into the hot water bath and cook for 30 minutes.
- Remove swordfish from the bag and pat dry with paper towel.
- Heat grill pan over high heat. Add swordfish and sear for 2 minutes on each side.
- Serve and enjoy.

Per Serving: Calories: 384; Total Fat: 22g; Saturated Fat: 4.4g; Protein: 43.2g; Carbs: 0g; Fiber: 0g; Sugar: 0g;

Bacon Wrapped Scallops

Serves: 4 / Preparation time: 10 minutes / Cooking time: 30 minutes / Temperature: 125 F / 52 C

8 scallops

1 tbsp shallots, chopped

8 bacon slices

3 fresh thyme sprigs

- Wrap each scallop with one bacons slice.
- Place scallop into the large zip-lock bag with shallots and thyme.
- Remove all air from the bag before sealing.
- Fill and preheat sous vide water oven to 125 F/ 52 C.
- Place bag in hot water bath and cook for 30 minutes.
- Remove scallops from the bag and pat dry with paper towel.
- Sear scallops on a grill pan until lightly brown.
- Serve and enjoy.

Per Serving: Calories: 260; Total Fat: 16.3g; Saturated Fat: 5.3g; Protein: 24.2g; Carbs: 2.4g; Fiber: 0g; Sugar: 0g;

Honey Ginger Salmon

Serves: 2 / Preparation time: 5 minutes / Cooking time: 30 minutes / Temperature: 125 F / 52 C

2 salmon fillets

2 tsp ginger, grated

3 tbsp tamari

2 tsp honey

1/8 tsp salt

- Fill and preheat sous vide water oven to 125 F/ 52 C.
- Add all ingredients to the large zip-lock bag and mix well.
- Remove all air from the bag before sealing.
- Place bag in hot water bath and cook for 30 minutes.
- Remove salmon from bag and pat dry with paper towel.
- Sear salmon skin side down until nicely brown.
- Serve and enjoy.

Per Serving: Calories: 279; Total Fat: 11g; Saturated Fat: 1.6g; Protein: 37.6g; Carbs: 8.5g; Fiber: 0.5g; Sugar: 6.3g;

Poached Lobster

Serves: 2 / Preparation time: 5 minutes / Cooking time: 20 minutes / Temperature: 140 F / 60 C

2 lobster tails

1 fresh tarragon sprig

6 tbsp butter, cut into pieces

Sea salt

- Fill and preheat sous vide water oven to 140 F/ 60 C.
- Add lobster tails into the large zip-lock bag.
- Add butter, tarragon, and salt into the bag.
- Remove all air from the bag before sealing.
- Place bag in hot water bath and cook for 20 minutes.
- Serve warm and enjoy.

Per Serving: Calories: 345; Total Fat: 34g; Saturated Fat: 21g; Protein: 0.4g; Carbs: 0.2g; Fiber: 0g; Sugar: 0g;

Mahi Mahi with Bean Puree

Serves: 2 / Preparation time: 10 minutes / Cooking time: 20 minutes / Temperature: 135 F / 57 C

2 mahi-mahi fillets

1 cup white wine

1/4 cup onion, chopped

1/4 cup squid ink

1 cup black beans, cooked

2 tbsp olive oil

Salt

- Fill and preheat sous vide water oven to 135 F/ 57 C.
- Season mahi-mahi fillets with salt and set aside for 5 minutes.
- Add fish fillets and olive oil into the zip-lock bag and remove all air from the bag before sealing.
- Place bag into the hot water bath and cook for 20 minutes.
- Meanwhile, sauté onion and black bean in a pan over medium heat.
- Pour in squid ink and white wine and simmer until reduced.
- Add black bean mixture to the blender and puree until smooth.
- Serve cooked mahi-mahi fish fillets on top of bean puree.

Per Serving: Calories: 662; Total Fat: 15.5g; Saturated Fat: 2.4g; Protein: 42.9g; Carbs: 68.5g; Fiber: 15.2g; Sugar: 3.7g;

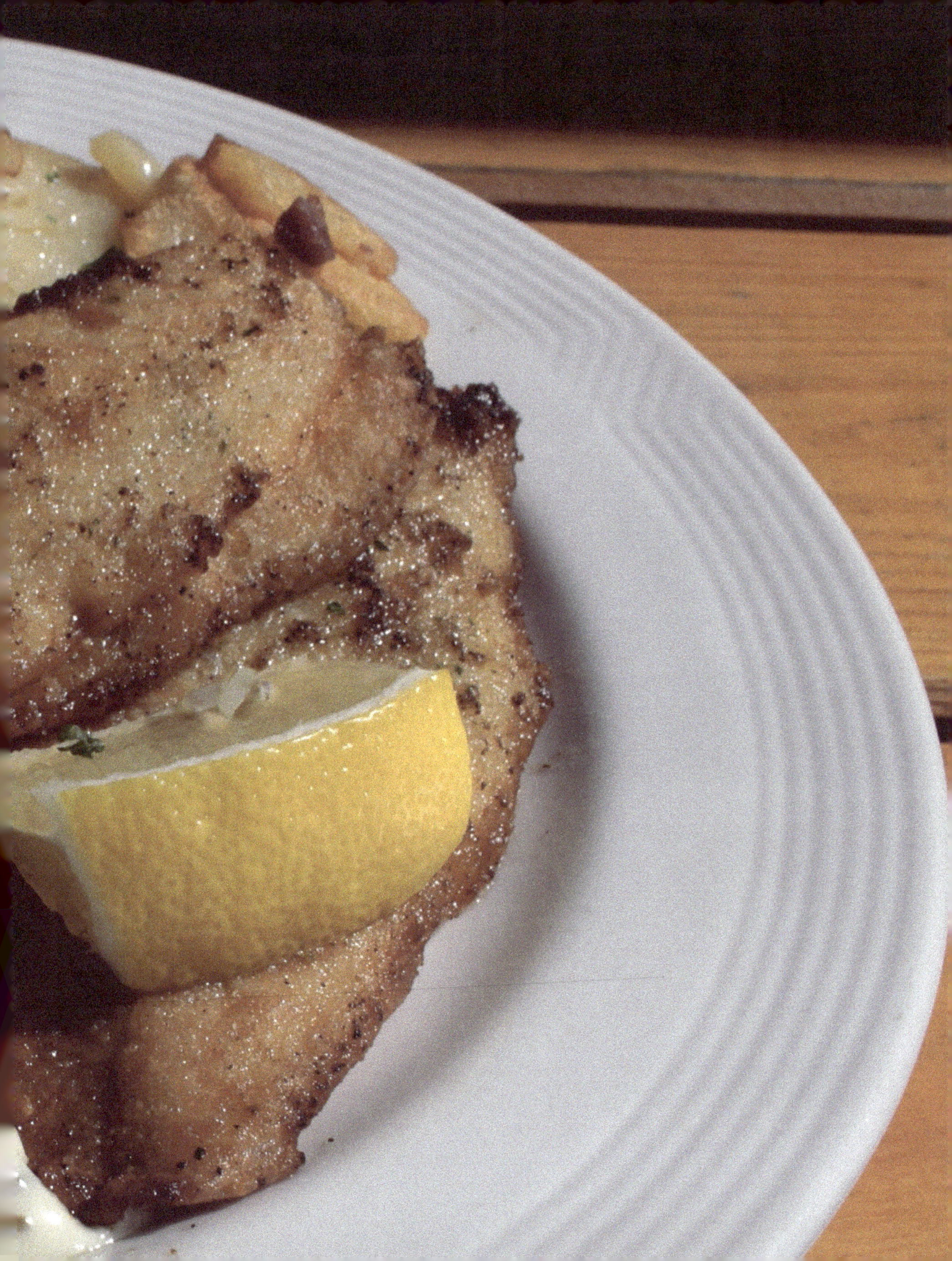

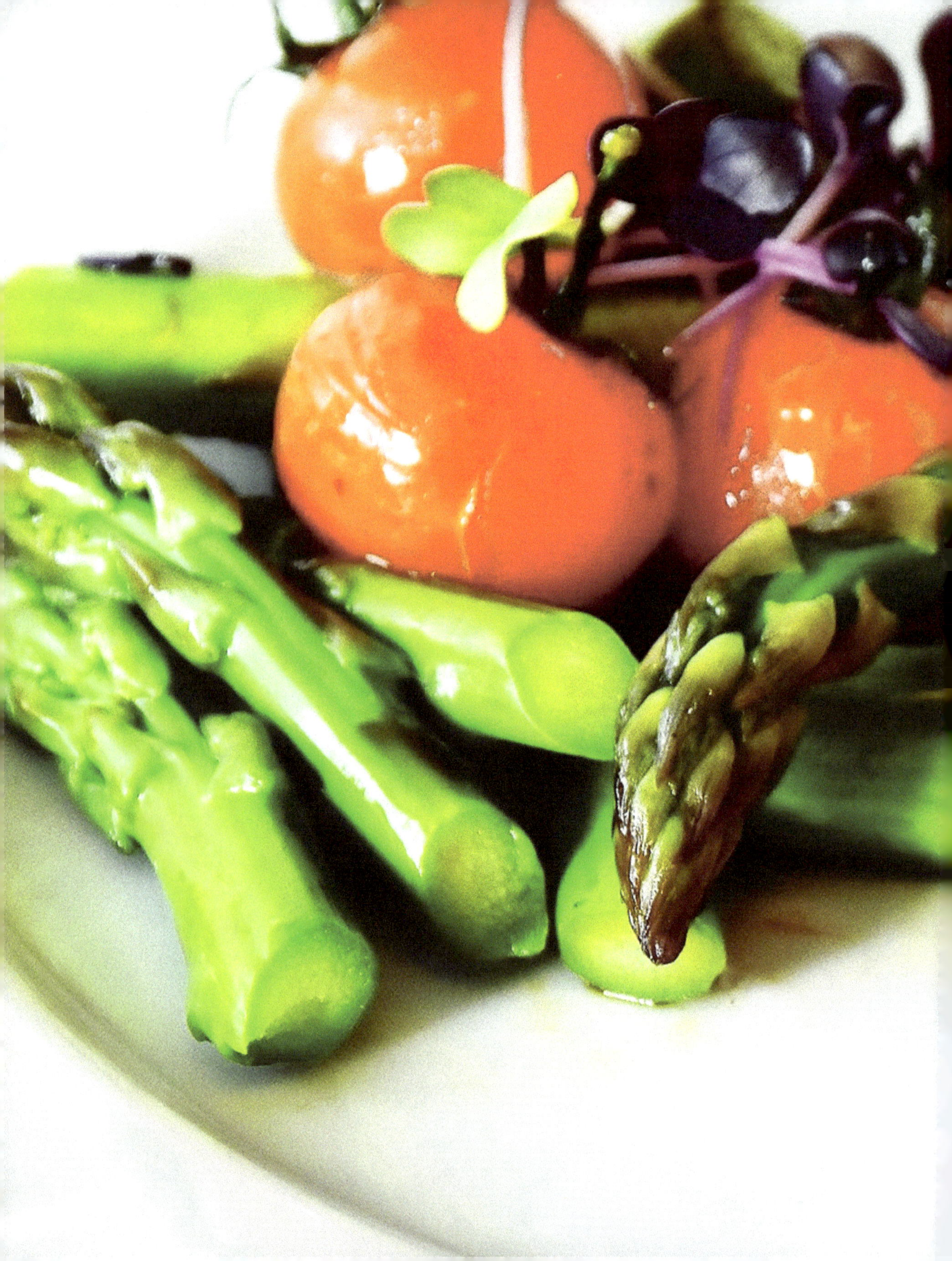

Chapter 6: Vegetable Recipes

Mashed Potatoes 107

Sous Vide Mushrooms 109

Drunken Garlic Onions 110

Garlicky Brussels Sprout 111

Poached Potatoes 112

Green Bean Casserole 113

Healthy Asparagus 114

Sweet and Spicy Carrots 115

Cauliflower Alfredo 116

Spicy Eggplant 117

Maple Sweet Potatoes 118

Creamy Cauliflower Puree 119

Creamy Spring Onion Soup 120

Garlic Risotto 121

Curried Carrots 122

Mashed Potatoes

Serves: 4 / Preparation time: 10 minutes / Cooking time: 1 hour 30 minutes / Temperature: 194 F / 90 C

32 oz russet potatoes, rinsed, peel, and sliced

5 garlic cloves smash

1 cup whole milk

3 rosemary springs

2 tbsp kosher salt

8 oz butter

- Preheat water oven to 194 F / 90 C.
- Place the potatoes into a zip-lock bag with butter, rosemary, garlic, milk, and salt. Remove all the air from the bag before sealing.
- Place bag into the hot water bath for 1 1/2 hours.
- Transfer potatoes into the bowl and mash until creamy. Stir in butter.
- Serve and enjoy.

Per Serving: Calories: 605; Total Fat: 48g; Saturated Fat: 30g; Protein: 6.5g; Carbs: 39g; Fiber: 5.5g; Sugar: 5.9g;

Sous Vide Mushrooms

Serves: 4 / Preparation time: 10 minutes / Cooking time: 30 minutes / Temperature: 176 F / 80 C

16 oz mushrooms, clean and cut into pieces

1/2 tsp black pepper

1 tbsp balsamic vinegar

2 tbsp soy sauce, low sodium

2 tbsp olive oil

2 tsp fresh thyme leaves

1/2 tsp kosher salt

- Preheat the water oven to 176 F / 80 C.
- Add all ingredients to the bowl and toss well.
- Transfer mushroom mixture into the zip-lock bag and remove all the air from the bag before sealing.
- Place bag into the hot water bath and cook for 30 minutes.
- Serve and enjoy.

Per Serving: Calories: 91; Total Fat: 7.4g; Saturated Fat: 1g; Protein: 4g; Carbs: 4.9g; Fiber: 1.5g; Sugar: 2.1g;

Drunken Garlic Onions

Serves: 1 / Preparation time: 5 minutes / Cooking time: 2 hours / Temperature: 185 F / 85 C

2 onions, sliced

2 garlic cloves, minced

1 tbsp olive oil

1 cup beer, lager

Pepper

Salt

- Preheat water oven to 185 F / 85 C.
- Add 1 tbsp olive oil in a pan and heat over medium heat.
- Add garlic and onion in pan, season with salt and pepper and sauté for 5 minutes.
- Add beer reduce heat at medium to low and stir for 5 minutes.
- Pour onion mixture into the zip-lock bag and remove all the air from the bag before sealing.
- Place bag into the hot water bath and cook for 2 hours.
- Transfer mixture into pan and cook over medium heat until liquid is reduced.
- Serve and enjoy.

Per Serving: Calories: 319; Total Fat: 14g; Saturated Fat: 2g; Protein: 3.9g; Carbs: 31g; Fiber: 4.9g; Sugar: 9.4g;

Garlicky Brussels Sprout

Serves: 6 / Preparation time: 10 minutes / Cooking time: 1 hour / Temperature: 185 F / 85 C

6 cup Brussels sprouts, halved

1 tbsp Tabasco sauce

1 cup butter, cubed

5 garlic cloves, chopped coarsely

3/4 tbsp kosher salt

- Preheat water oven to 185 F / 85 C.
- Add all ingredients to the bowl and toss well.
- Transfer bowl mixture into the zip-lock bag and remove all the air from the bag before sealing.
- Place bag into the hot water bath and cook for 1 hour.
- Remove the bag from the water bath and chill quickly into a bowl of ice water.
- Refrigerated until 2 hours before serving.
- Serve and enjoy.

Per Serving: Calories: 313; Total Fat: 31g; Saturated Fat: 19g; Protein: 3.5g; Carbs: 8.9g; Fiber: 3.4g; Sugar: 2g;

Poached Potatoes

Serves: 6 / Preparation time: 10 minutes / Cooking time: 1 hour / Temperature: 190 F / 87 C

16 oz small potatoes, cut in half

1 tbsp fresh thyme, minced

1 tbsp olive oil

2 tbsp butter

2 tbsp kosher salt

1 tbsp black pepper

- Preheat water oven to 190 F / 87 C.
- Combine all ingredients into a zip-lock bag and remove all the air from the bag before sealing.
- Place the bag into the hot water bath and cook for 1 hour.
- Serve and enjoy.

Per Serving: Calories: 110; Total Fat: 6.3g; Saturated Fat: 2.8g; Protein: 1.5g; Carbs: 12.9g; Fiber: 2.3g; Sugar: 0.9g;

Green Bean Casserole

Serves: 3 / Preparation time: 10 minutes / Cooking time: 2 hours / Temperature: 185 F / 85 C

2 cups fresh green beans, trimmed and cut into pieces

7 oz mushroom soup, condensed cream

1 egg, beaten

1 can crispy fried onion

1/2 medium shallot, peeled and diced

1/2 cup breadcrumbs

Pepper

Salt

- Preheat the water oven to 185 F / 85 C.
- Mix all ingredients except crispy onion, into a large bowl and toss well to combine.
- Transfer green bean mixture into the zip-lock bag and remove all the air from the bag before sealing.
- Place the bag into a water bath and cook for 2 hours.
- Remove the bag from the water bath and spoon the mixture into serving plate and top with fried onion.
- Serve and enjoy.

Per Serving: Calories: 171; Total Fat: 6.4g; Saturated Fat: 1.6g; Protein: 6.7g; Carbs: 22.8g; Fiber: 3.3g; Sugar: 3.2g;

Healthy Asparagus

Serves: 3 / Preparation time: 5 minutes / Cooking time: 15 minutes / Temperature: 175 F / 79 C

16 oz of fresh asparagus, trimmed

1 lemon zest

3 tbsp olive oil

Pepper

Salt

- Preheat the water oven to 175 F / 79 C.
- Put the asparagus in a zip-lock bag with olive oil, lemon zest, salt, and pepper.
- Remove all the air from the bag before sealing.
- Place bag into a water bath and cook for 15 minutes.
- Remove the bag from the water bath and transfer asparagus to a serving plate.
- Serve and enjoy.

Per Serving: Calories: 150; Total Fat: 14.2g; Saturated Fat: 2.1g; Protein: 3.3g; Carbs: 5.9g; Fiber: 3.2g; Sugar: 2.8g;

Sweet and Spicy Carrots

Serves: 4 / Preparation time: 10 minutes / Cooking time: 1 hour / Temperature: 183 F / 83 C

- 4 large carrots, peeled and chopped into 1/2 inches pieces
- 1 tbsp honey
- 1 tbsp butter
- 1 tsp white vinegar
- 1 tsp sweet paprika
- 1/2 tsp salt
- 1/2 tsp chili powder
- 4 tbsp chopped parsley

- Preheat water oven to 183 F / 83 C.
- Combine all ingredients in zip-lock bag except parsley.
- Remove all the air from the bag before sealing.
- Place bag into the hot water bath and cook for 1 hour.
- Remove the bag from the water bath and transfer carrots on a serving dish and garnish with parsley.
- Serve and enjoy.

Per Serving: Calories: 74; Total Fat: 3g; Saturated Fat: 1.8g; Protein: 0.8g; Carbs: 12g; Fiber: 2.1g; Sugar: 7.9g;

Cauliflower Alfredo

Serves: 2 / Preparation time: 10 minutes / Cooking time: 2 hours / Temperature: 181 F / 83 C

2 cup cauliflower florets, chopped

1/2 cup vegetable stock

2 tbsp milk

2 tbsp butter

2 garlic cloves, crushed

Pepper

Salt

- Preheat the water oven to 181 F / 83 C.
- Put all ingredients into a zip-lock bag and remove all the air from the bag before sealing.
- Place bag into the hot water bath and cook for 2 hours.
- Remove the bag from the water bath and transfer mixture into a blender and blend until smooth.
- Season with salt and pepper.
- Serve and enjoy.

Per Serving: Calories: 142; Total Fat: 12.5g; Saturated Fat: 8g; Protein: 2.8g; Carbs: 7.6g; Fiber: 2.6g; Sugar: 3.6g;

Spicy Eggplant

Serves: 4 / Preparation time: 10 minutes / Cooking time: 45 minutes / Temperature: 185 F / 85 C

4 thai eggplants, cut into wedges

1 tbsp sesame seeds

1 tbsp brown sugar

2 tbsp soy sauce

2 tbsp deonjang paste

1/4 cup peanut oil

- Fill and preheat sous vide water oven to 185 F/ 85 C.
- In a bowl, whisk together deonjang paste, sugar, soy sauce, and peanut oil.
- Add eggplants to the bowl and toss well.
- Transfer eggplant mixture into the zip-lock bag and remove all the air from the bag before sealing.
- Place bag into the hot water bath and cook for 45 minutes.
- Drain eggplant from liquid and sear in a hot pan.
- Garnish with sesame seeds and serve.

Per Serving: Calories: 170; Total Fat: 14.6g; Saturated Fat: 2.4g; Protein: 1.9g; Carbs: 8.4g; Fiber: 2.3g; Sugar: 5.3g;

Maple Sweet Potatoes

Serves: 2 / Preparation time: 10 minutes / Cooking time: 1 hour 30 minutes / Temperature: 175 F / 79 C

1 large sweet potato, cut into chunks

2 tbsp water

4 tbsp maple syrup

4 tbsp butter

- Fill and preheat sous vide water oven to 175 F/ 79 C.
- Add sweet potato chunks, water, half butter, and half maple syrup into the zip-lock bag and remove all the air from the bag before sealing.
- Place bag into the hot water bath and cook for 1 1/2 hours.
- Heat remaining butter and maple syrup into the pan over medium heat.
- Transfer sweet potato chunks into the pan and stir until well coated.
- Serve and enjoy.

Per Serving: Calories: 389; Total Fat: 23g; Saturated Fat: 14.6g; Protein: 2.1g; Carbs: 45.5g; Fiber: 3g; Sugar: 29.7g;

Creamy Cauliflower Puree

Serves: 6 / Preparation time: 10 minutes / Cooking time: 45 minutes / Temperature: 180 F / 82 C

1 medium cauliflower head, cut into florets

2 oz heavy cream

2 tbsp butter

Pepper

Salt

- Fill and preheat sous vide water oven to 180 F/ 82 C.
- Add all ingredients to the zip-lock bag and remove all the air from the bag before sealing.
- Place bag into the hot water bath and cook for 45 minutes.
- Transfer bag contents to the bowl and using blender puree until smooth.
- Serve warm and enjoy.

Per Serving: Calories: 91; Total Fat: 7.4g; Saturated Fat: 4.6g; Protein: 2.1g; Carbs: 5.4g; Fiber: 2.4g; Sugar: 2.3g;

Creamy Spring Onion Soup

Serves: 2 / Preparation time: 10 minutes / Cooking time: 45 minutes / Temperature: 180 F / 82 C

1 cup spring onions, rinsed and trimmed

1 cup water

1 tsp olive oil

1 tsp soy sauce

2 garlic cloves, chopped

1 medium potato, peeled and diced

Black pepper

Salt

- Fill and preheat sous vide water oven to 180 F/ 82 C.
- Add all ingredients to the large zip-lock bag and remove all the air from the bag before sealing.
- Place bag into the hot water bath and cook for 45 minutes.
- Transfer bag mixture to the blender and puree until smooth.
- Season with pepper and salt.
- Serve and enjoy.

Per Serving: Calories: 124; Total Fat: 2.6g; Saturated Fat: 0.4g; Protein: 3.4g; Carbs: 23.5g; Fiber: 3.7g; Sugar: 2.1g;

Garlic Risotto

Serves: 4 / Preparation time: 10 minutes / Cooking time: 45 minutes / Temperature: 183 F / 83 C

1 cup Arborio rice

1/3 cup Romano cheese, grated

1 tsp rosemary, minced

3 cups vegetable broth

2 tbsp garlic, minced

1 tsp olive oil

Pepper

Salt

- Fill and preheat sous vide water oven to 183 F/ 83 C.
- Add all ingredients except cheese into the zip-lock bag and remove all the air from the bag before sealing.
- Place bag into the hot water bath and cook for 45 minutes.
- Pour rice mixture into the bowl.
- Add cheese to the risotto and stir well.
- Serve and enjoy.

Per Serving: Calories: 244; Total Fat: 4.5g; Saturated Fat: 1.9g; Protein: 9.7g; Carbs: 40.1g; Fiber: 1.5g; Sugar: 0.6g;

Curried Carrots

Serves: 4 / Preparation time: 5 minutes / Cooking time: 45 minutes / Temperature: 183 F / 84 C

5 carrots, peeled and sliced

1/2 tsp curry powder

1 tbsp butter

Black pepper

Salt

- Fill and preheat sous vide water oven to 183 F/ 84 C.
- Add all ingredients to the zip-lock bag and remove all the air from the bag before sealing.
- Place bag into the hot water bath and cook for 45 minutes.
- Remove carrots from bag and place on serving the dish.
- Serve and enjoy.

Per Serving: Calories: 58; Total Fat: 2.9g; Saturated Fat: 1.8g; Protein: 0.7g; Carbs: 7.7g; Fiber: 2g; Sugar: 3.8g;

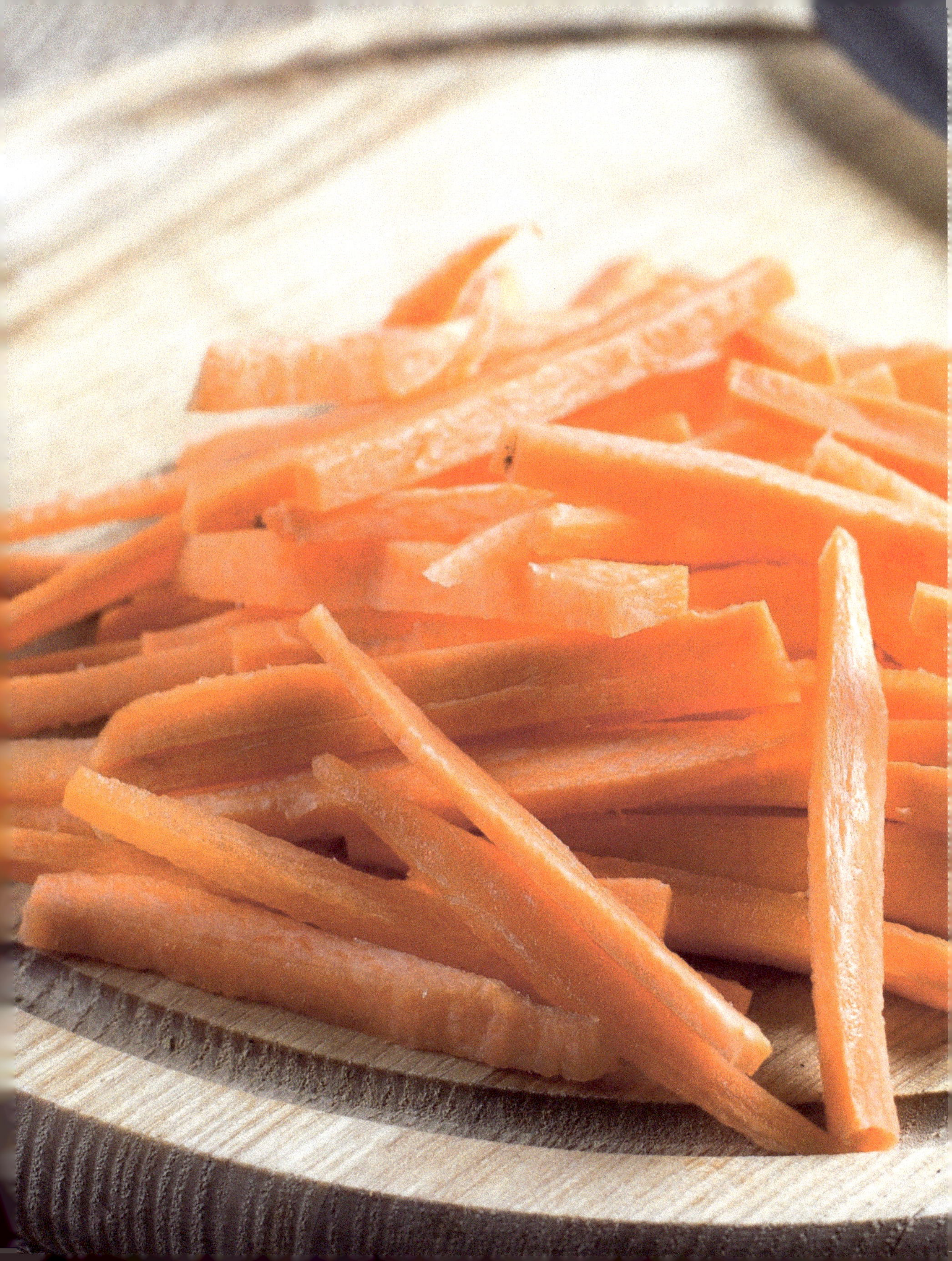

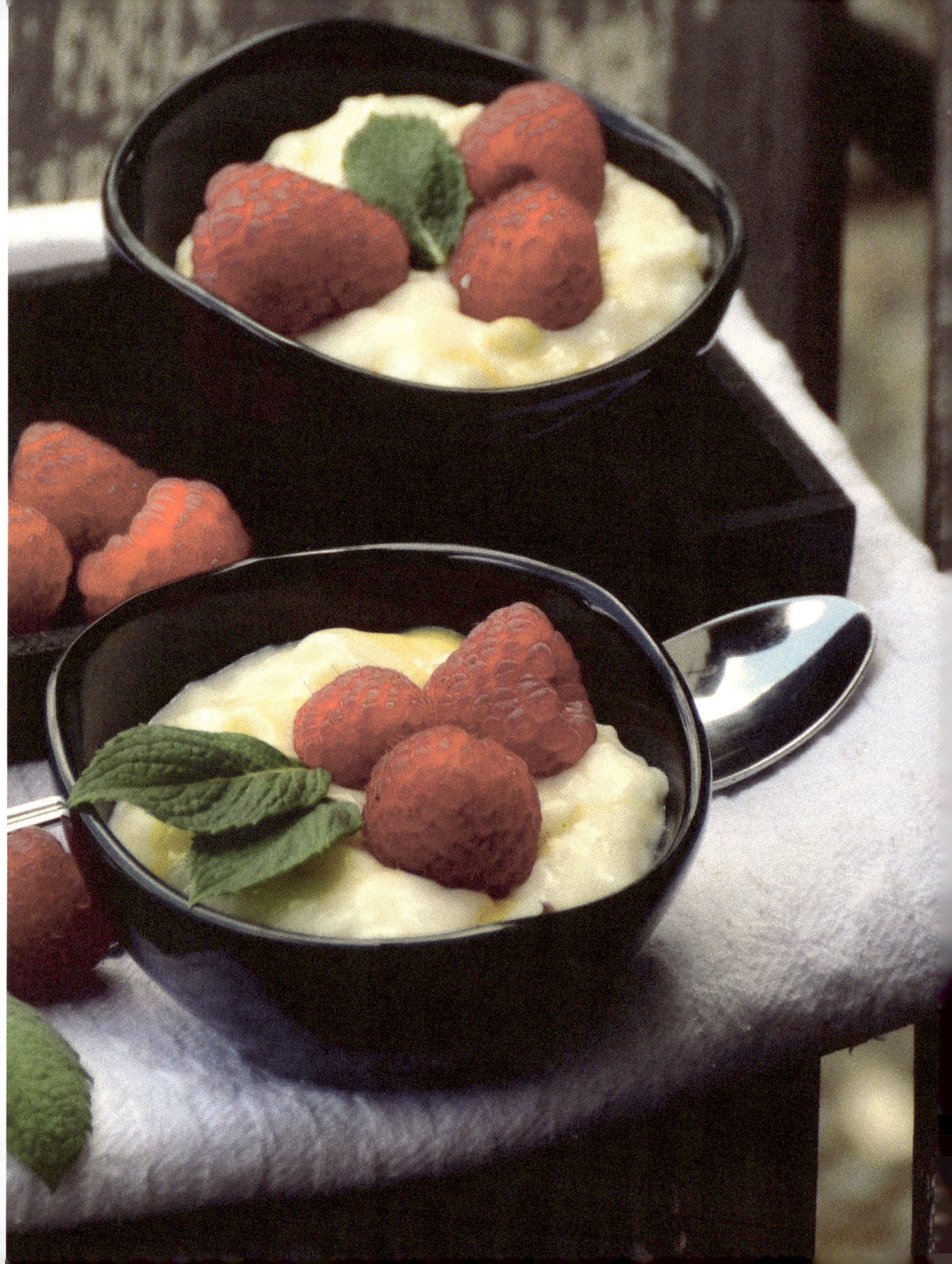

Chapter 7: Dessert Recipes

Sweet Poached Pears 126

Maple Apples 127

Cinnamon Apples 128

Doce de Banana 129

Simple Applesauce 130

Sous Vide Peaches 131

Raisin Rice Pudding 132

Vanilla Pudding 133

Easy Sweet Kiwi 134

Coconut Rice Pudding 135

Sweet Poached Pears

Serves: 4 / Preparation time: 10 minutes / Cooking time: 1 hour / Temperature: 175 F / 79 C

4 ripe pears, peeled

1 tsp vanilla extract

1 tsp orange zest

1/4 cup sweet vermouth

1/2 cup granulated sugar

1 cup red wine

1 tsp salt

- Fill and preheat sous vide water oven to 175 F/ 79 C.
- Add all ingredients to the zip-lock bag and remove all the air from the bag before sealing.
- Place bag into the hot water bath and cook for 1 hour.
- Remove pears from the bag. Core and slice pears.
- Serve with ice cream and enjoy.

Per Serving: Calories: 267; Total Fat: 0.3g; Saturated Fat: 0g; Protein: 0.8g; Carbs: 58.7g; Fiber: 6.5g; Sugar: 46g;

Maple Apples

Serves: 3 / Preparation time: 10 minutes / Cooking time: 4 hours / Temperature: 190 F / 87 C

3 apples, peeled, cored and sliced

1 tbsp maple whiskey

1 tbsp sugar

1/8 tsp salt

- Add all ingredients to the zip-lock bag and remove all air from the bag before sealing.
- Fill and preheat sous vide water oven to 190 F/ 87 C.
- Place bag into the hot water bath and cook for 4 hours.
- Serve warm and enjoy.

Per Serving: Calories: 131; Total Fat: 0.4g; Saturated Fat: 0g; Protein: 0.6g; Carbs: 34.8g; Fiber: 5.4g; Sugar: 27.2g;

Cinnamon Apples

Serves: 4 / Preparation time: 5 minutes / Cooking time: 30 minutes / Temperature: 185 F / 85 C

4 medium apples, peeled and sliced

1/4 tsp ground cinnamon

1 tbsp brown sugar

2 tbsp butter

- Fill and preheat sous vide water oven to 185 F/ 85 C.
- Add all ingredients to the zip-lock bag and remove all air from the bag before sealing.
- Place bag in hot water bath and cook for 30 minutes.
- Serve and enjoy.

Per Serving: Calories: 176; Total Fat: 6.2g; Saturated Fat: 3.7g; Protein: 0.7g; Carbs: 33.1g; Fiber: 5.5g; Sugar: 25.4g;

Doce de Banana

Serves: 4 / Preparation time: 10 minutes / Cooking time: 30 minutes / Temperature: 175 F / 80 C

5 small bananas, peeled and cut into chunks

6 whole cloves

2 cinnamon sticks

1 cup brown sugar

- Fill and preheat sous vide water oven to 176 F/ 80 C.
- Add all ingredients to the zip-lock bag and remove all air from the bag before sealing.
- Place bag into the hot water bath and cook for 30 minutes.
- Remove from bag and allow to cool slightly.
- Discard cloves and cinnamon sticks.
- Top with ice cream and serve.

Per Serving: Calories: 250; Total Fat: 0.4g; Saturated Fat: 0.1g; Protein: 1.4g; Carbs: 64.4g; Fiber: 3.3g; Sugar: 50.6g;

Simple Applesauce

Serves: 4 / Preparation time: 5 minutes / Cooking time: 45 minutes / Temperature: 180 F / 82 C

4 apples, peeled, cored and chopped

1/2 tsp ground cinnamon

1/4 cup sugar

1 lemon juice

- Fill and preheat sous vide water oven to 180 F/ 82 C.
- Add all ingredients to the zip-lock bag and remove all air from the bag before sealing.
- Place bag into the hot water bath and cook for 45 minutes.
- Transfer apple mixture to the blender and blend until getting desired consistency.
- Serve warm and enjoy.

Per Serving: Calories: 164; Total Fat: 0.4g; Saturated Fat: 0g; Protein: 0.6g; Carbs: 43.5g; Fiber: 5.6g; Sugar: 35.7g;

Sous Vide Peaches

Serves: 4 / Preparation time: 10 minutes / Cooking time: 30 minutes / Temperature: 180 F / 82 C

4 peaches, halved and pitted

1/3 cup brandy

1/2 cup sugar

1/2 cup water

- Fill and preheat sous vide water oven to 180 F/ 82 C.
- Heat sugar and water into the saucepan. Once sugar is dissolved then remove the pan from heat.
- Stir in brandy and set aside.
- Add peach halved into the zip-lock bag then pour syrup over the peaches.
- Remove all air from the bag before sealing.
- Place bag into the hot water bath and cook for 30 minutes.
- Serve and enjoy.

Per Serving: Calories: 153; Total Fat: 0.4g; Saturated Fat: 0g; Protein: 1.4g; Carbs: 39g; Fiber: 2.3g; Sugar: 39g;

Raisin Rice Pudding

Serves: 6 / Preparation time: 10 minutes / Cooking time: 2 hours / Temperature: 140 F / 60 C

2 cups Arborio rice

1/2 tsp ground ginger

2 tsp ground cinnamon

1/2 cup maple syrup

1/2 cup raisins

2 tbsp butter

3 cups milk

- Fill and preheat sous vide water oven to 140 F/ 60 C.
- Add all ingredients to the zip-lock bag and stir well.
- Remove all air from the bag before sealing.
- Place bag into the hot water bath and cook for 2 hours.
- Stir well and serve warm.

Per Serving: Calories: 430; Total Fat: 6.8g; Saturated Fat: 4g; Protein: 8.7g; Carbs: 84.3g; Fiber: 2.6g; Sugar: 28.3g;

Vanilla Pudding

Serves: 6 / Preparation time: 10 minutes / Cooking time: 45 minutes / Temperature: 180 F / 82 C

1 cup milk

1 tbsp vanilla extract

3 tbsp cornstarch

3 large eggs

2 large egg yolks

1/2 cup sugar

1 cup heavy cream

1/8 tsp salt

- Fill and preheat sous vide water oven to 180 F/ 82 C.
- Add all ingredients in the blender and puree until smooth.
- Pour pudding mixture into the zip-lock bag and seal the bag.
- Place bag into the hot water bath and cook for 45 minutes.
- Once it done then remove from water bath and pour bag contents into the blender and blend until smooth.
- Pour into the bowl and place in refrigerator.
- Serve chilled and enjoy. Garnish with your favorite berries.

Per Serving: Calories: 227; Total Fat: 12.2g; Saturated Fat: 6.4g; Protein: 5.8g; Carbs: 23.5g; Fiber: 0g; Sugar: 19g;

Easy Sweet Kiwi

Serves: 2 / Preparation time: 10 minutes / Cooking time: 20 hours / Temperature: 176 F / 80 C

2 kiwis, peeled and sliced

2 tbsp sugar

1 tbsp lemon juice

- Fill and preheat sous vide water oven to 176 F/ 80 C.
- Add all ingredients to the zip-lock bag and remove all air from the bag before sealing.
- Place bag into the hot water bath and cook for 20 minutes.
- Serve and enjoy.

Per Serving: Calories: 93; Total Fat: 0.5g; Saturated Fat: 0.1g; Protein: 0.9g; Carbs: 23.3g; Fiber: 2.3g; Sugar: 19g;

Coconut Rice Pudding

Serves: 6 / Preparation time: 10 minutes / Cooking time: 4 hours / Temperature: 180 F / 82 C

1/2 cup Arborio rice

1/2 tsp ground ginger

1 tsp ground cinnamon

1/2 cup coconut, shredded

1/2 cup sugar

14 oz can coconut milk

2 cups milk

- Fill and preheat sous vide water oven to 180 F/ 82 C.
- Add all ingredients to the zip-lock bag and remove all air from the bag before sealing.
- Place bag into the hot water bath and cook for 4 hours.
- Stir well and serve warm.

Per Serving: Calories: 316; Total Fat: 18.1g; Saturated Fat: 15.5g; Protein: 5.3g; Carbs: 36.6g; Fiber: 1.3g; Sugar: 20.8g;

The "Dirty Dozen" and "Clean 15"

Every year, the Environmental Working Group releases a list of the produce with the most pesticide residue (Dirty Dozen) and a list of the ones with the least chance of having residue (Clean 15). It's based on analysis from the U.S. Department of Agriculture Pesticide Data Program report.

The Environmental Working Group found that 70% of the 48 types of produce tested had residues of at least one type of pesticide. In total there were 178 different pesticides and pesticide breakdown products. This residue can stay on veggies and fruit even after they are washed and peeled. All pesticides are toxic to humans and consuming them can cause damage to the nervous system, reproductive system, cancer, a weakened immune system, and more. Women who are pregnant can expose their unborn children to toxins through their diet, and continued exposure to pesticides can affect their development.

This info can help you choose the best fruits and veggies, as well as which ones you should always try to buy organic.

The Dirty Dozen

Strawberries
Spinach
Nectarines
Apples
Peaches
Celery

Grapes
Pears
Cherries
Tomatoes
Sweet bell peppers
Potatoes

The Clean 15

Sweet corn
Avocados
Pineapples
Cabbage
Onions
Frozen sweet peas
Papayas
Asparagus

Mangoes
Eggplant
Honeydew
Kiwi
Cantaloupe
Cauliflower
Grapefruit

Measurement Conversion Tables

Volume Equivalents (Liquid)

US Standard	US Standard (ounces)	Metric (Approx.)
2 tablespoons	1 fl oz	30 ml
¼ cup	2 fl oz	60 ml
½ cup	4 fl oz	120 ml
1 cup	8 fl oz	240 ml
1 ½ cups	12 fl oz	355 ml
2 cups or 1 pint	16 fl oz	475 ml
4 cups or 1 quart	32 fl oz	1 L
1 gallon	128 fl oz	4 L

Oven Temperatures

Fahrenheit (F)	Celsius (C) (Approx)
250°F	120°C
300°F	150°C
325°F	165°C
350°F	180°C
375°F	190°C
400°F	200°C
425°F	220°C
450°F	230°C

Volume Equivalents (Dry)

US Standard	Metric (Approx.)
¼ teaspoon	1 ml
½ teaspoon	2 ml
1 teaspoon	5 ml
1 tablespoon	15 ml
¼ cup	59 ml
½ cup	118 ml
1 cup	235 ml

Weight Equivalents

US Standard	Metric (Approx.)
½ ounce	15 g
1 ounce	30 g
2 ounces	60 g
4 ounces	115 g
8 ounces	225 g
12 ounces	340 g
16 ounces or 1 pound	455 g

Cooking Times and Temperatures

The following is meant to be a general guide to cooking times and temperatures for your Anova Sous Vide. You can use this as a quick reference when you are creating your own recipes, or want to change things up by improvising or substituting ingredients for the recipes found in this cookbook. This isn't meant to be a definitive guide, as the appropriate time and temperature can vary according to the type and cut of meat as well as the specific recipe you are cooking. Predictably, lower temperatures require longer cooking times – sometimes much longer. Don't be afraid to experiment! With this guide at your fingertips, you'll get a great dish every time.

Chicken

	Temperature	Min (hours)	Max (hours)
Soft, juicy breast	145°F / 63°C	1:30	4:00
Traditional (firm) breast	155°F / 69°C	1:00	4:00
Juicy, tender thigh	165°F / 74°C	1:00	4:00
Off-the-bone tender thigh	165°F / 74°C	4:00	8:00

Finish cooking chicken by drying the meat thoroughly and then cooking it with the skin side down in a skillet set to medium heat with a drizzle of your preferred cooking oil.

Steak or Lamb

	Temperature	Min (hours)	Max (hours)
Medium-rare	129°F / 54°C	1:30	4:00
Medium	140°F / 60°C	1:30	4:00
Medium-well	145°F / 63°C	1:30	4:00

Finish cooking steaks by drying the meat thoroughly and then searing it quickly. Either a hot cast iron skillet drizzled with your preferred oil, or a hot grill will work well for searing the meat.

Pork Chops

	Temperature	Min (hours)	Max (hours)
Rosy pink, juicy	135°F / 57°C	1:00	4:00
White throughout, juicy	147°F / 64°C	1:00	4:00
Medium-well	145°F / 63°C	1:30	4:00

Finish cooking pork chops by drying the meat thoroughly and then searing it quickly. Either a hot cast iron skillet drizzled with your preferred oil, or a hot grill will work well for searing the meat.

Fish

	Temperature	Min (hours)	Max (hours)
"Mi-cuit" (tender, translucent)	110°F / 43°C	0:30	1:00
Tender and flaky	124°F / 51°C	0:30	1:00
Traditional well-done	132°F / 55°C	0:30	1:30

It is a challenge to sear sous vide cooked fish without causing it to fall apart. Sometimes it is a trade-off between searing the fish and having the perfect presentation.

Eggs

	Temperature	Min (hours)	Max (hours)
Soft yolks and barely set whites	140°F / 60°C	0:55	1:05
Creamy yolks that are opaque in appearance with tender whites	145°F / 63°C	0:45	1:05

To make your poached eggs turn out perfectly, you will want to cook them for 45 minutes at 145°F, then remove them from their shells. Next, you poach them carefully for 30 seconds in simmering water. If you want soft boiled eggs, try cooking them for 3 minutes in boiling water. Next, transfer the eggs to a 145°F water bath for 45 minutes. Leave their shells on for the water bath.

Sausage

	Temperature	Min (hours)	Max (hours)
Extra-juicy, soft	140°F / 60°C	0:45	4:00
Firm, juicy	150°F / 66°C	0:45	4:00
Traditional	160°F / 71°C	0:45	4:00

Sausages can be finished on a grill, or in a skillet drizzled with your preferred cooking oil.

Hamburger

	Temperature	Min (hours)	Max (hours)
Very Rare (aka "Blue Rare") to Rare	115-123°F / 46-51°C	0:40	2:30
Medium-Rare	124-129°F / 51-54°C	0:40	2:30
Medium	130-137°F / 54-58°C	0:40	4:00
Medium-Well	138-144°F / 59-62°C	0:40	4:00
Well-Done	145-155°F / 63-68°C	0:40	4:00

Green Vegetables

	Temperature	Min (hours)	Max (hours)
Asparagus, broccoli, peas, etc.	183°F / 84°C	0:15	0:40

Root Vegetables

	Temperature	Min (hours)	Max (hours)
Carrots, potatoes, parsnips, etc.	183°F / 84°C	1:00	3:00

Recipe Index

B
Bacon Wrapped Scallops 101
Balsamic Chicken Breast 58
Barbacoa Tacos 30
BBQ Baby Back Ribs 81
BBQ Cranberry Salmon 96
BBQ Pork Ribs 79
Beef Brisket 19
Beef Meatballs 29
Beef Roast 21
Beef Tenderloin with Butter 26
Bone- In Pork Chop 76
Buttery Halibut 94

C
Cauliflower Alfredo 116
Chicken Adobo 65
Chicken Ancho Chile 63
Chicken Breast with Lemon Sauce 61
Chicken Fajitas 62
Chicken Legs 66
Chicken Meatballs 64
Chicken Thighs 55
Chicken Wings 59
Chicken with Tomato Vinaigrette 56
Creamy Cauliflower Puree 119
Creamy Spring Onion Soup 120
Crispy Skin Salmon 98
Curried Carrots 122

D
Drunken Garlic Onions 110

E
Easy Pecan Salmon 92

F
Filet Mignon with Chimichurri 25
Five Spice Pork 80
Flank Steak 24
Flavorful Bavette Steak 32
Fried Chicken 60

G
Garlic Butter Lamb Chops 38
Garlic Risotto 121
Garlicky Brussels Sprout 111
Greek Chicken Meatballs 67
Green Bean Casserole 113

H
Healthy Asparagus 114
Herb Butter Garlic Steaks 18
Herb Garlic Lamb Chops 40
Herb Rub Pork Chops 75
Honey Ginger Salmon 102

J
Juicy and Tender Swordfish 100

L
Lamb Burgers 42
Lamb Chops with Basil Chimichurri 37
Lamb Loin with Mint Olive Salsa 44
Lamb Rack with Herb Butter 41
Lamb Steaks 49
Lamb with Mint Gremolata 46
Lemon Pork Chops 78
Lemon Thyme Chicken 57

M
Mahi Mahi with Bean Puree 104
Maple Sweet Potatoes 118
Mashed Potatoes 108
Meatballs 22
Meatballs with Sauce 50

P
Perfect Pork Chop 85
Poached Lobster 103
Poached Potatoes 112
Pork Carnitas 82
Pork Chops with Mushrooms 73
Pork Loin 77
Pork Tenderloin 74
Pulled Pork 83

R
Rib Eye Steak 20
Rosemary Garlic Lamb Chops 36

S
Scallops with Salsa Verde 91
Simple Brown Butter Scallops 95
Simple Chicken Breast 54
Simple Flank Steak 28
Simple No Sear Chicken Breast 68
Simple Rack of Lamb 39
Simple Sliced Pork Belly 84
Simple Spice Lamb 43
Smoked Paprika Shrimp 90
Sous Vide Cod 97
Sous Vide Mushrooms 109
Soy Lemon Lamb Rack 48
Spicy Eggplant 117
Steak Strips with Herbs 23
Steak with Chimichurri Sauce 27
Sweet and Spicy Carrots 115
Sweet and Spicy Pork Ribs 86

T
Tasty Teriyaki Salmon 99
Tender and Juicy Beef Brisket 31
Tender Lamb Chops 47
Tender Pork Chops 72
Thyme Rosemary Lamb 45
Tuna 93

Complete Index

Contents 5
Introduction 6
    What is Sous Vide? 6
    How does the Sous Vide cooking method work? 6
    Is Sous Vide Safe? 8
    Does cooking food Sous Vide lead to better meals? 9
    How do Sous Vide Ovens / Baths usually work? 10
    How does the Anova Precision Cooker differ from conventional Sous Vide machines? 10
    Using the Anova Sous Vide Precision Cooker 11
    Tips and tricks for using the Anova Sous Vide Precision Cooker: 11
    How do I take care of my Anova Sous Vide Precision Cooker? 13

Chapter 1: Beef Recipes 17
Herb Butter Garlic Steaks 18
Beef Brisket 19
Rib Eye Steak 20
Beef Roast 21
Meatballs 22
Steak Strips with Herbs 23
Flank Steak 24
Steak with Chimichurri 25
Beef Tenderloin with Butter 26
Steak with Chimichurri Sauce 27
Simple Flank Steak 28
Beef Meatballs 29
Barbacoa Tacos 30
Tender and Juicy Beef Brisket 31
Flavorful Bavette Steak 32

Chapter 2: Lamb Recipe 35
Rosemary Garlic Lamb Chops 35

Lamb Chops with Basil Chimichurri 37
Garlic Butter Lamb Chops 38
Simple Rack of Lamb 39
Herb Garlic Lamb Chops 40
Lamb Rack with Herb Butter 41
Lamb Burgers 42
Simple Spice Lamb 43
Lamb Loin with Mint Olive Salsa 44
Thyme Rosemary Lamb 45
Lamb with Mint Gremolata 46
Tender Lamb Chops 47
Soy Lemon Lamb Rack 48
Lamb Steaks 49
Meatballs with Sauce 50

Chapter 3: Chicken Recipe 53
Simple Chicken Breast 54
Chicken Thighs 55
Chicken with Tomato Vinaigrette 56
Lemon Thyme Chicken 57
Balsamic Chicken Breast 58
Chicken Wings 59
Fried Chicken 60
Chicken Breast with Lemon Sauce 61
Chicken Fajitas 62
Chicken Ancho Chile 63
Chicken Meatballs 64
Chicken Adobo 65
Chicken Legs 66
Greek Chicken Meatballs 67
Simple No Sear Chicken Breast 68

Chapter 4: Pork Recipes 71
Tender Pork Chops 72
Pork Chops with Mushrooms 73
Pork Tenderloin 74

Anova Sous Vide Cookbook

Herb Rub Pork Chops 75
Bone- In Pork Chop 76
Pork Loin 77
Lemon Pork Chops 78
BBQ Pork Ribs 79
Five Spice Pork 80
BBQ Baby Back Ribs 81
Pork Carnitas 82
Pulled Pork 83
Simple Sliced Pork Belly 84
Perfect Pork Chop 85
Sweet and Spicy Pork Ribs 86

Chapter 5: Seafood Recipes 89
Smoked Paprika Shrimp 89
Scallops with Salsa Verde 91
Easy Pecan Salmon 92
Tuna 93
Buttery Halibut 94
Simple Brown Butter Scallops 95
BBQ Cranberry Salmon 96
Sous Vide Cod 97
Crispy Skin Salmon 98
Tasty Teriyaki Salmon 99
Juicy and Tender Swordfish 100
Bacon Wrapped Scallops 101
Honey Ginger Salmon 102
Poached Lobster 103
Mahi Mahi with Bean Puree 104

Chapter 6: Vegetable Recipes 107
Mashed Potatoes 107
Sous Vide Mushrooms 109
Drunken Garlic Onions 110
Garlicky Brussels Sprout 111
Poached Potatoes 112

Green Bean Casserole 113
Healthy Asparagus 114
Sweet and Spicy Carrots 115
Cauliflower Alfredo 116
Spicy Eggplant 117
Maple Sweet Potatoes 118
Creamy Cauliflower Puree 119
Creamy Spring Onion Soup 120
Garlic Risotto 121
Curried Carrots 122

Chapter 7: Desserts Recipe 125
Sweet Poached Pears 126
Maple Apples 127
Cinnamon Apples 128
Doce de Banana 129
Simple Applesauce 130
Sous Vide Peaches 131
Raisin Rice Pudding 132
Vanilla Pudding 133
Easy Sweet Kiwi 134
Coconut Rice Pudding 135

The "Dirty Dozen" and "Clean 15" 136
    The Dirty Dozen 136
    The Clean 15 136
Measurement Conversion Tables 137
Cooking Times and Temperatures 138
  Chicken 138
  Steak or Lamb 138
  Pork Chops 139
  Fish 139
  Eggs 139
  Sausage 140
  Hamburger 140
  Green Vegetables 140

Root Vegetables 140
Recipe Index 141
Complete Index 143

Want MORE healthy recipes for FREE?

Double down on healthy living with a full week of fresh, healthy salad recipes. A new salad for every day of the week!

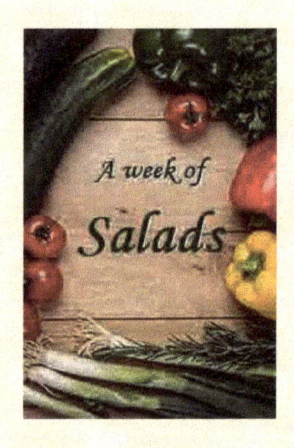

Grab this bonus recipe ebook *free* as our gift to you:

http://salad7.hotbooks.org

Want MORE full length cookbooks for FREE?

We invite you to sign up for free review copies of future books!

Learn more and get brand new cookbooks for *free*:

http://club.hotbooks.org

CPSIA information can be obtained
at www.ICGtesting.com
Printed in the USA
LVHW070049241020
669611LV00001BA/10